Be Seated *by* Bemis

A 100-YEAR HISTORY OF BEMIS MANUFACTURING COMPANY

Bemis Project Team: Richard A. Bemis, Kari A. Bramstedt, Margaret S. Harper, Karen E. Hoefler

Concept, research, writing, design and layout:
Jacob & Clark Marketing Communications, Inc., Sheboygan, Wisconsin
Project director and copy editor: William G. Jacob
Research and copywriting: Denise M. Pazur
Design and layout: Debra J. Coopman

Additional historical research: Jan Hildebrand, Sheboygan Falls, Wisconsin; and Peter Laun, Sheboygan, Wisconsin

Published by: Times Printing Co. Inc., Random Lake, Wisconsin
Binding by: Inland Press/Inland Book, Menomonee Falls, Wisconsin

Library of Congress Control Number: 2001130649

ISBN: 0-9711791-0-7

First printing: July 2001

Trademarks:
The following are current or former trademarks of Bemis Manufacturing Company and its subsidiaries: ADF™, Bemis Health Care®, Bemis Casual Living™, The Best Seat in the House®, Born in Barrie Raised Everywhere®, Carrara & Matta™, Cavalier®, Colonial Kitchen®, Church®, Décor®, Festival®, Lawn Comfort®, Mayfair®, Molcor®, Moldex®, Posturemold®, Proseat®, Rain Master®, Tommer®, Top-Tite®, UltraHinge™, The Visible Difference in Resin Furniture®,Waterwick®

Photography Credits:
Select photos in Chapters 1-4 courtesy of The Peter Laun Collection, housed and preserved at the Sheboygan County Historical Research Center, Sheboygan Falls, Wisconsin

All other pictures courtesy of the Bemis family, Bemis Manufacturing Company, employees and friends

Photography of Bemis artifacts by Blue Moon Studio, Inc., Sheboygan, Wisconsin
Photography of Bemis Board of Directors and executive management by Image Studios, Appleton, Wisconsin

Image of historic poster of "Rosie the Riveter" from Corbis Images; Elvis Presley, *The Graduate* poster, LBJ swearing-in, and Vietnam War images from Getty Images

Founding family members of Bemis Manufacturing Company, circa 1925, included Albert Bemis (left) with son F.K. "Pete" Bemis (right). Joining Al and Pete are Al's wife Oleida (second from right) and a family friend. In 1925 Al purchased a majority interest in White Wagon Works of Sheboygan Falls, Wisconsin. Together, Al and Pete, along with Pete's sons Dick and Peter, have built Bemis into the world-class company it is today.

TABLE OF CONTENTS

A simple child's coaster wagon formed the foundation of Bemis Manufacturing Company. At the turn of the twentieth century, the company's precursor, White Wagon Works of Sheboygan Falls, Wisconsin, crafted thousands of wooden coaster wagons with a unique steering invention that set them apart from the competition. Shown at right are two early workers at White Wagon Works, surrounded by wooden wagon bodies, wheels and frames.

DEDICATION

Perhaps it was the wagon. Or the industry. Or the era.

Or maybe it was just good fortune, combined with a hefty dose of courage and persistence, that has brought Bemis Manufacturing Company to its 100th year in business.

Whatever the reasons for the century of success that Bemis celebrates in 2001, one thing is certain—the company owes that success to its employees.

This book is dedicated to them.

Were it not for the thousands of men and women, past and present, who shared their ideas, ingenuity and inventions, Bemis would not be the robust and well-grounded company it is today. In offering their many talents, our employees have given not only to Bemis but to the industries we support and to the communities in which we reside.

As we consider where the next century will take Bemis Manufacturing Company, let's pause to reflect on where we have come from. *Be Seated by Bemis* records that rich history and gives us a glimpse of what our future may bring.

It is with gratitude and pride that we share this special story with you.

Richard A. Bemis
PRESIDENT
CHIEF EXECUTIVE OFFICER

Peter F. Bemis
EXECUTIVE VICE PRESIDENT
PRESIDENT, CONTRACT GROUP

AUGUST 2001

No. 663,526. Patented Dec. 11, 1900.

A. WHITE.

BOY'S WAGON.

(Application filed July 30, 1900.)

(No Model.)

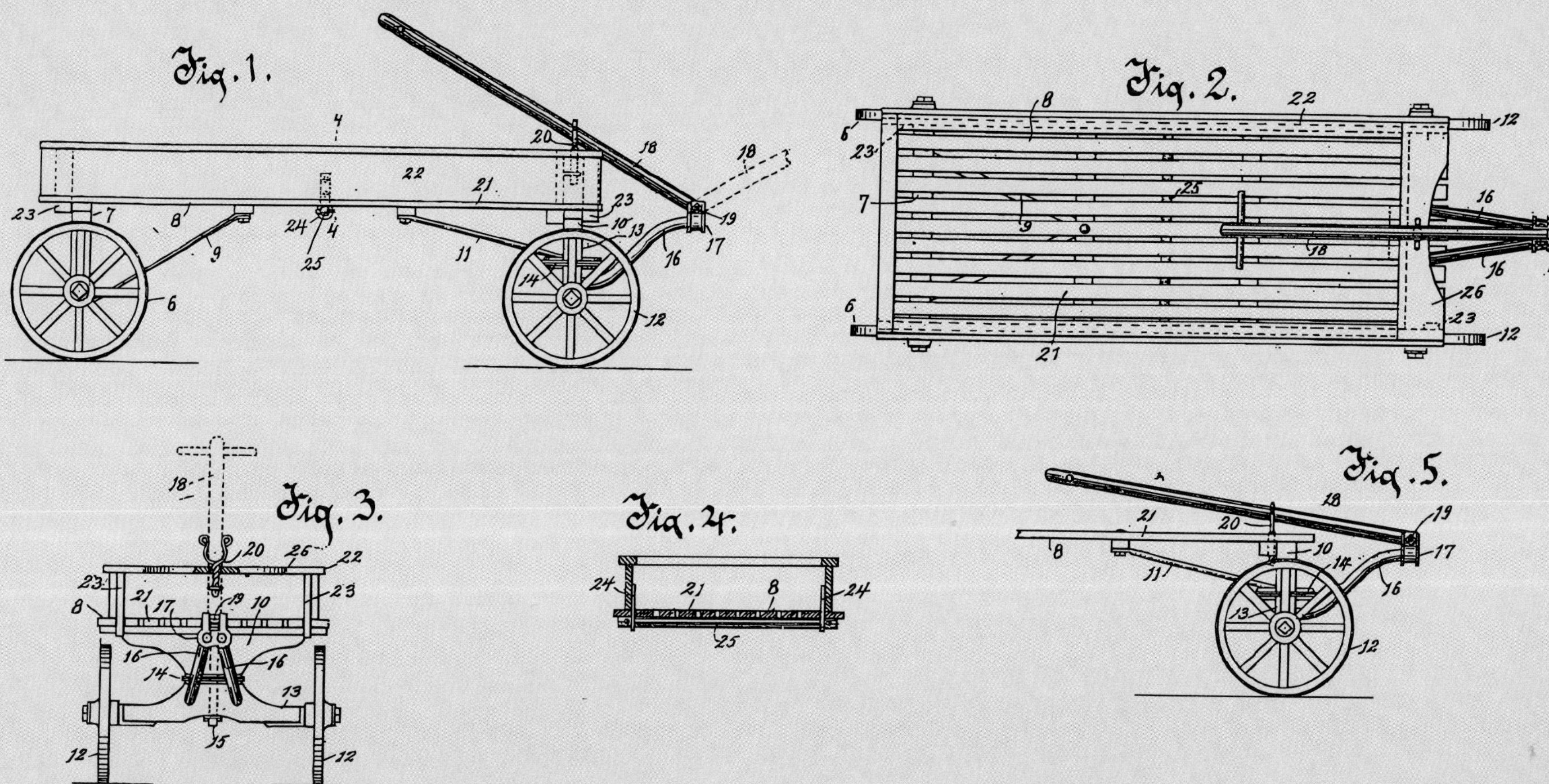

Witnesses.

Anna V. Faust

Inventor.

Arthur White

By Benedict & Morsell

Attorneys.

CHAPTER 1

It all began with a wagon

A child's wooden wagon. Simple as it is, this wagon—part toy, part transportation, part imagination—was where Bemis Manufacturing Company began.

It was the dawn of the twentieth century, and Sheboygan County, Wisconsin, was a growing industrial center surrounded by agricultural land and divided by waterways. Entrepreneurs and craftsmen from the East Coast settled in the county's namesake city of Sheboygan, drawn by the abundance of Wisconsin hardwood and the natural port on Lake Michigan.

In those days Sheboygan was known as "Chair City" because of the many furniture makers who set up shop there. Among them was an eccentric inventor named Arthur White.

Arthur had traveled to Sheboygan from his home in New Hampshire to peddle a rocker spring he had invented. But it was another of his devices—a simple steering mechanism for a child's wagon—that led him to Sheboygan Falls and formed the foundation of Arthur's most successful venture—White Wagon Works.

Original drawings of the White wagon design (opposite page). Inventor Arthur White received patents based on these sketches in 1900.

Workers of yesteryear

An old payroll register from 1907 shows that the typical worker at White Wagon Works toiled 10 hours a day and was paid $1.75 for that time, or about 18 cents an hour. Workers' names show evidence of long-standing families in the Sheboygan Falls area. They include:

Back, John
Chapman, H.
Fredricks, Ferd
Hemsehmeyer, Al
Hemsehmeyer, A.O.
Hemsehmeyer, C.H.
Hemsehmeyer, Fred
Kohlhagen, Arthur
Kohlhagen, Joe
Krichner, Frank
Krichner, G.
Meyer, Otto
Meyer, Paul
Mueller, L.
Schaap, Fred
Steffen, Arthur
Trowbridge, Frank
Voight, Emil
Wagner, Leo
Wilsey, Fred

Employees of White Wagon Works, circa 1914

Innovation—the legacy of Bemis Manufacturing Company

In 1900 Arthur White patented a new steering mechanism for a child's coaster wagon.

Arthur recognized that children's wagons were exact replicas of full-size wagons. But the solid front bar that was harnessed to horses on the adult variety made the child's version difficult to steer. So Arthur altered the steering system for his coaster wagons to make them more child-friendly.

He began by tilting the shaft back so that a youngster could sit or kneel in the wagon and steer at the same time. He also invented an interlocking ring mechanism so wagon wheels would rotate in response to driver's commands.

These modest inventions turned the White wagon into one of the most coveted toys of its day. They also formed the foundation of Bemis Manufacturing Company.

Birth of White Wagon Works

Arthur White had all he needed to launch a business—his patented steering mechanism for a child's wagon, two shareholders to raise $10,000 in capital and his dogged determination. With these he incorporated White Wagon Works on August 6, 1901.

A modest building on Walnut Street in Sheboygan Falls was his company's first location. Formerly a manufacturing site for pumps and milk safes, the building cost a mere $72 a year to rent.

White coaster wagons proved a fast success because of their novel steering mechanism, quality workmanship and versatile design feature that allowed removal of the wagon's sides.

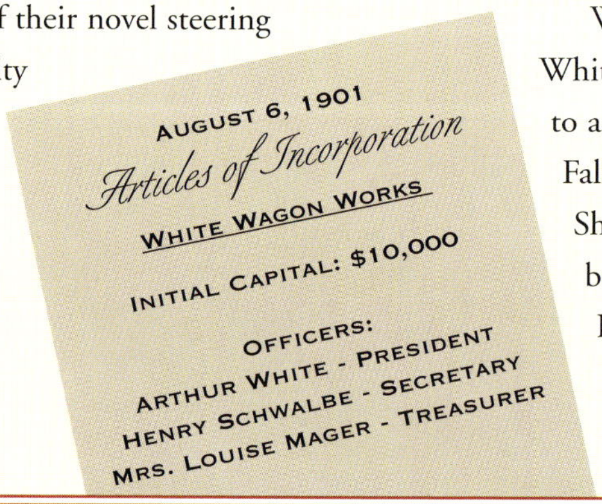

After just two years in business, White Wagon Works ran out of room at the Walnut Street facility. Arthur needed money to expand and so he searched for new shareholders. One investor was Sheboygan business owner Otis Trowbridge. A grocer by trade, Otis purchased one-quarter interest in White Wagon Works in 1903 and was named company secretary. His wife was Luella Bemis Trowbridge, and her brother Albert would later take a vital interest in White Wagon Works. Thus began a Trowbridge family relationship with Bemis that continues today.

With money from new investors, White Wagon Works relocated in 1903 to a building in downtown Sheboygan Falls near the upper dam of the Sheboygan River. In later years this building would come to be known as Bemis Plant A.

1901	1903	1912	1915	1917	1920	1924	1928
Arthur White incorporates White Wagon Works, the precursor to Bemis.	White Wagon Works relocates to a former mill, later known as Plant A.	The *RMS Titanic* sinks in the Atlantic Ocean on its maiden voyage.	Albert Bemis and George Riddell establish Bemis-Riddell Fibre Co.	The Uncle Sam "I Want You" recruiting poster is unveiled as the U.S. enters WW I.	The Chicago White Sox throw the World Series and are indicted for the crime of gambling.	Al Bemis and George Riddell buy the majority of shares of White Wagon Works.	Al Bemis takes controlling interest in White Coaster Wagon Works.

Wagon business booms

By the summer of 1908 wanderlust had struck Arthur White. He left the business that bore his name, selling his interest in White Wagon Works and leaving shareholders Otis Trowbridge and Abner Heald in charge.

Together they took White Wagon Works to new successes.

From 1908 to the early 1920s, sales climbed as customers from far-flung places including Baltimore, Kansas City, Louisville and even South Africa clamored for the company's popular children's coaster wagons. Newspaper publishers across the United States even purchased White wagons for their delivery boys.

Fueling sales further were detachable runners that converted White wagons to sleds. This allowed youngsters to enjoy their wagons year-round.

The growing popularity and increasing demand for the product of the White Wagon Works of this city, is something remarkable. The White wagons are on sale in nearly every village and city in the U.S. and Canada, and they have also been shipped to South Africa. Several carloads are shipped yearly to Washington D.C., and recently the Washington Times, one of the capitol City's greatest newspapers, ordered 75 of these wagons to be used by a part of their force of delivery boys. These wagons will bear the following legend on either side of the box: 'Washington Times Delivery Service. This will be a good ad for the White Wagon Works.

Local newspaper account from July 1919 shows how popular White wagons had become in the company's heyday.

As business boomed at the Wagon Works, staff scrambled to fill orders, especially at the holidays when Christmas orders poured in. It was common in those days for employees to work 12-hour shifts, and company owners enlisted the help of relatives to keep wagons rolling off production lines.

Otis and Luella Bemis Trowbridge with son Walter. Luella was the first Bemis family member to become affiliated with White Wagon Works, the precursor to Bemis Manufacturing Company.

An aerial view of Sheboygan Falls at the turn of the twentieth century. White Wagon Works, which later became Bemis Plant A, is visible in the center of this photograph.

F.K. "Pete" Bemis, circa 1925

One of the family members who helped out during the 1919 Christmas rush was Francis von Kaas Bemis, or "Pete," as he liked to be called. Just 9 years old at the time, Pete would later lead Bemis Manufacturing Company to great success. The nephew of Wagon Works owners Otis and Luella Trowbridge, Pete also pitched in at his father Albert's business, the Bemis-Riddell Fibre Co.

All about Albert

If any one person in the richly woven history of Bemis Manufacturing Company can be called its founder, that person is Albert Bemis.

Born June 28, 1877, in Sheboygan Falls, Wisconsin, Al was the son of Frank and Gertrude Dennison Bemis. He was a business-minded man with a desire to succeed in Sheboygan's flourishing furniture industry. After attending Sheboygan Business College, Al joined Crocker Chair Company in 1897 at age 20. There he worked for a former business

Albert Bemis and Oleida—a love story

"Albert 'darling' Bemis was over . . . and we talked and talked. My brain is in a perfect turmoil at times because of so much thinking."
(TAKEN FROM THE DIARY OF OLEIDA VON KAAS BEMIS)

Oleida von Kaas often wrote about beau Albert and their three-year courtship in her journals—including concerns that Al was "a pretty sensible boy" but had "an exalted opinion of me, however."

As their courtship continued, Al and Oleida took carriage rides and had long talks about the future. Soon they were seeing each other almost every day.

The couple was wed on June 13, 1903, at the von Kaas home in Sheboygan, Wisconsin. Oleida's ring cost a mere $8.46, and the marriage license 75 cents. A rented horse and rig were just $2.

Al and Oleida had three children: Gertrude, born in 1904; Francis von Kaas, or "Pete," born in 1910; and Rebecca, born in 1921. They were married for 43 years when Oleida died in 1943. Al passed away just three years later.

associate of his late father Frank. In the 1860s Frank Bemis and Crocker Chair had partnered to form the Bemis Bros. & Crocker Chair Company. Since that time Crocker had grown to become a thriving Sheboygan-area business that employed 1,000 people during its heyday in the late 1890s.

Al shared his wages earned as a shipping clerk at Crocker Chair with family members, purchasing groceries for his widowed mother, paying her utility bills and giving spending money to his younger brother Charles. But he found more than just a job and steady income at Crocker Chair Company. Al also found a wife.

Dark-haired and striking in appearance, Oleida von Kaas was the daughter of Hans von Kaas, who oversaw the paint shop at Crocker Chair and was a friend of Al's parents.

Al and Oleida first became acquainted on a sleigh ride in March 1900. Their courtship revolved around dancing at the local yacht club and landau rides through Sheboygan. Their marriage followed three years later in 1903.

Al stayed at Crocker Chair Company for 17 years in all. During that time he worked his way up to superintendent. In 1914 Al accepted a position as assistant superintendent at American Manufacturing Company, a rival area furniture maker.

American was manufacturing a new product called fibre furniture, and these wicker-like chairs, tables and carriages were to become Al's specialty.

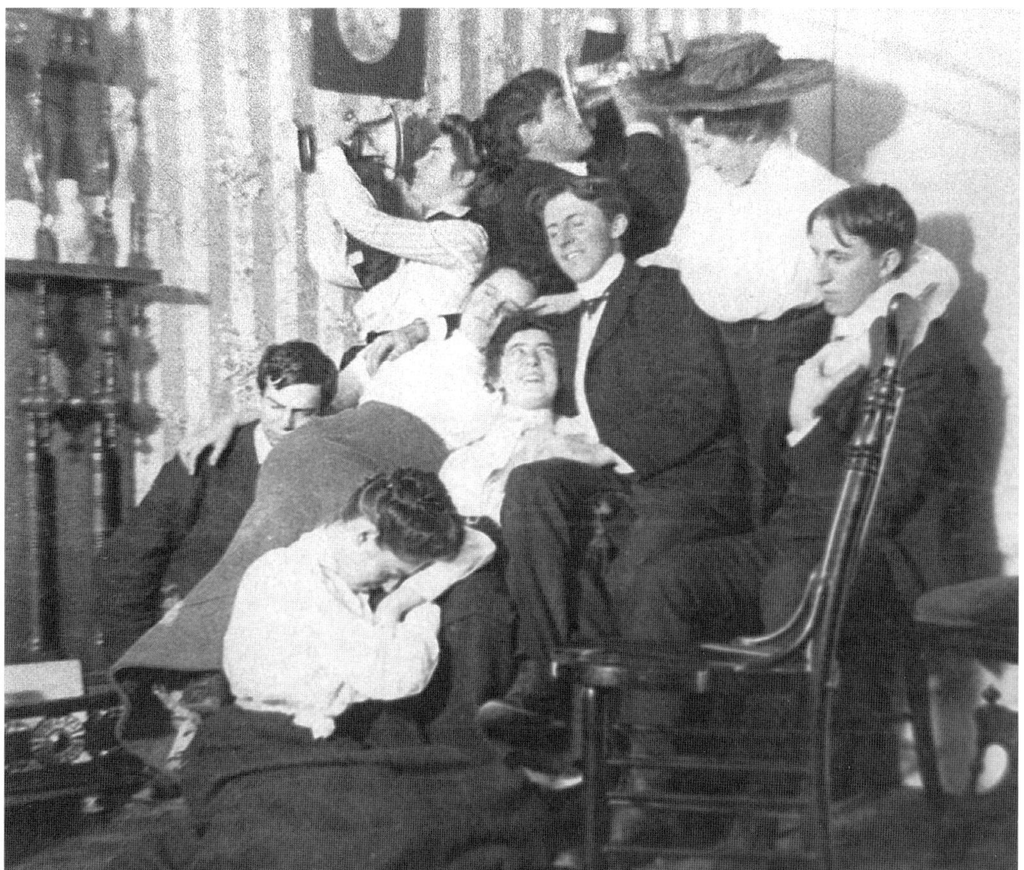

Bemis Manufacturing Company founder Al Bemis (center seated) with wife Oleida (lower front) and friends, circa 1902. A Sheboygan area furniture maker, Al purchased an interest in White Wagon Works in 1924 and began to shape that modest firm into its successor, Bemis Manufacturing Company.

Fibre furniture fever

In his new position at American Manufacturing Company, Al Bemis saw how successful fibre furniture was. That inspired him to strike out on his own.

So in 1915, just one year after he joined American Manufacturing, Al partnered with colleague George Riddell to form their own company, Bemis-Riddell Fibre Co. of Sheboygan. It was one of just a handful of companies that fed the nation's fever for fibre furniture.

BEMIS-RIDDELL FIBRE CO.
SHEBOYGAN, WIS.

A new design trend, the popular wicker-like woven paper furniture was costly to manufacture, inexpensively priced and the rage of the 1920s. Part of America's infatuation with fibre furniture can be linked to the growing use of automobiles. Front porches became a popular place to sit and watch the cars go by, and fibre furniture was in vogue for porches.

As it gained in popularity, fibre furniture even found acceptance indoors. The home of Kohler Co. founder John Michael Kohler, located on New York Avenue and Sixth Street in Sheboygan, included pieces of fibre furniture when it was donated to the American Legion in 1923.

Fibre furniture reigned in popularity until newer designs, more cost-effective manufacturing processes and the Great Depression cooled interest and sales. But while it was a hot trend, fibre furniture set the country's interests on fire. Bemis-Riddell Fibre Co. fanned the flames—and prospered along the way.

Advertisements for Bemis-Riddell Fibre Co., circa 1920, promoted the many uses for the firm's diverse fibre products.

Oleida Bemis, wife of company founder Al Bemis, holding her youngest child Rebecca in 1921. Oleida is rocking her infant daughter on a chair manufactured by her husband's firm, Bemis-Riddell Fibre Co. of Sheboygan.

Furniture making—a Bemis heritage

Albert Bemis wasn't the first family member to develop a passion for furniture making. He likely inherited it from his father Frank.

Francis Tower Bemis was a pioneer in Sheboygan's late-nineteenth-century furniture industry. He came to Wisconsin from Massachusetts in 1868 and a year later founded the Bemis Bros. & Crocker Chair Company with brother Herbert and another local furniture manufacturer. When the partnership ended after just a year, Frank and Herbert moved west to Sheboygan Falls to start their own business.

The Bemis Brothers Chair Company was a successful enterprise for seven years until fire struck in 1877. Frank and Herbert rebuilt a smaller factory and continued their business until 1884, when it was destroyed by another fire.

Frank persisted in making furniture on his own until his death six years later. Albert was just 13 years old when his father died in 1890, but a love of furniture making and an ironclad will to succeed had already taken root in him.

Furniture-making brothers Francis Tower Bemis (left) and Herbert Bemis, circa 1870

Walter Trowbridge, circa 1915

Decline of White Wagon Works

In 1924 production at Sheboygan's Bemis-Riddell Fibre Co. was rising, and Al Bemis and partner George Riddell were reaping the benefits.

But business at White Wagon Works in Sheboygan Falls had taken a downturn. The company was reeling after the death of Otis Trowbridge in 1921. Vice president and general manager of the Wagon Works, Otis was also brother-in-law to Al Bemis.

When Otis died, his wife Luella and son Walter assumed leadership roles in the business, but by then sales were in a slump. Dozens of North American manufacturers were making products patterned after the White wagon when its design patent expired in 1917. Youngsters preferred the competitors' wagons because they featured quiet rubber tires and sleek metal bodies, instead of the old-fashioned White wooden wagons. Competitors were also diversifying, offering other items including toys and pony carts.

> ***"Papa & I tried to sell wagons and it's heck too—but I guess they'll sell alright later on . . ."***
>
> –LETTER FROM OLEIDA BEMIS TO DAUGHTER GERTRUDE, 1925
> *(Written from the Chicago Furniture Mart, at an exhibit where Al and Oleida showcased their coaster wagons)*

Trouble at the Wagon Works was further compounded in 1924 when shareholder Abner Heald announced retirement and the desire to sell his portion of the business. New company Vice President Luella Bemis Trowbridge likely tipped off her brother Al regarding the pending change in ownership status at White Wagon Works.

Al was not one to pass up a business opportunity. So in December 1924 Bemis-Riddell Fibre Co. acquired the majority share in White Wagon Works for $20,000.

Al renamed the company White Coaster Wagon Works, and he and George became officers in the firm. Day-to-day operations were left to Luella and her son Walter.

In 1925 Al founded Bemis Manufacturing Company to act as a sales organization for White Coaster Wagon Works. BMC sold goods manufactured at the Wagon Works to its customers.

Entrepreneur and Bemis Manufacturing Company founder Al Bemis (second from right) with his family, (from left) daughter Gertrude, wife Oleida, daughter Rebecca and son "Pete," circa 1925. Al had recently purchased an interest in his sister Luella's firm, White Wagon Works, to add to his ownership of Bemis-Riddell Fibre Co.

Distinctive designs of smokers Stands and Cabinets are shown on the following pages. Pieces with a range in both style and price to please the most critical customer as well as adapt themselves to the one that is limited in means.

The trays and tray sets used are carefully chosen to conform with the design of the piece and are useful as well as decorative.

No. 627
Size of top 9 x 15 in.
Height overall 29 in.
Material—Red Gum
Finish—Mahogany, Walnut or Shaded Colors
Packed in individual cartons
Weight 7 lbs.

No. 504
Size of top 10 x 10 in.
Height overall 29½ in.
Material—5 ply Mahogany or Walnut top and doors.
3 ply Mahogany or Walnut panels.
Red Gum turnings.
Finish—Antique Mahogany, Walnut or Shaded Colors.
Packed in individual cartons
Weight 10 lbs.

No. 628
Size of top 9 x 15 in.
Height overall 30 in.
Material—Red Gum
Finish—Mahogany, Walnut or Shaded Colors.
Packed in individual cartons
Weight 7 lbs.

***Toys, radio cabinets and smoking sets**—all were added to the product mix at White Coaster Wagon Works during the mid- to late-1920s to boost lagging sales and diversify the company's offerings.*

Company President Al Bemis even brought a craftsman onboard—Elijah Pierson Wright—from his other business venture, Bemis-Riddell Fibre Co. Known as "Speed" because of his glib tongue, Wright created end tables, wall shelves, magazine racks and other novelties that were sold under the Bemis name. But the economy was changing, and company sales continued to decline.

Al goes his own way

Al Bemis and George Riddell maintained partial ownership of White Coaster Wagon Works from 1924 to 1928, while continuing to juggle management of their namesake firm, Bemis-Riddell Fibre Co. During those years each gravitated toward the business that most interested him—Al to the Wagon Works and George to the Fibre Co.

In 1928 they parted ways. George sold his shares of the Wagon Works to Al, and Al sold his shares of Bemis-Riddell Fibre Co. to George. It was a wise business decision for Al because the fibre furniture business declared bankruptcy just six years later.

In 1929, as the Great Depression gripped the country, Al's Wagon Works enterprise was not immune. Many people couldn't afford novelty furniture or children's toys, so company sales continued to fall. Al knew his wood-based manufacturing business was at a crossroads. So he turned his consideration to a more essential everyday product that was made of wood—toilet seats.

Al Bemis (right) with his only son, Francis von Kaas, known as "Pete." As Al made decisions about the future of his businesses, White Coaster Wagon Works and its sales organization, Bemis Manufacturing Company, son Pete was old enough to watch and learn. Pete would eventually take a leadership role at his family's company.

In the late 1920s Bemis Manufacturing Company produced sales flyers like this one to spotlight its line of novelty furniture. Manufactured by White Coaster Wagon Works and marketed by Bemis, the novelty furniture was an attempt to diversify the Wagon Works' product line and boost flagging sales.

Longtime Bemis employee Hank Back poses with a Bemis delivery truck, circa 1938. During those years the company's loading dock was located on the Sheboygan River at the old Broadway Avenue plant, later known as Plant A.

CHAPTER 2

How toilet seats saved Bemis

As the shadow of the Great Depression fell over the country, banks, shops and factories from coast to coast closed their doors.

It was the early 1930s and soup lines were a more common sight than paychecks. Businesses had just two choices as the nation grew more desperate—press on as they had in the past or adapt to the changing business climate. Al Bemis chose to adapt—and faced the challenges that came with it. His company was forced to lay off workers to cope with declining sales. In 1931 Al had 23 employees, but a year later that number dropped to 10.

In the early 1930s, Al Bemis took a chance on a new product line—toilet seats.

Many of Al's contemporaries were having similar problems during the sparse Depression years. These included the once-dominant Crocker Chair Company of Sheboygan where Al began his career more than 30 years earlier. When Crocker declared bankruptcy in 1932, Al recognized a business opportunity for Bemis.

DUMB-BELLS INCREASE THE STRENGTH Interlocked Seat CROCKER CHAIR CO. ANTIGO, WISCONSIN. PATENT APPLIED FOR

Bemis bought patents for the Crocker Interlocked Seat in 1932.

A new product—toilet seats

Demand for White coaster wagons was waning, and novelty furniture was a luxury few Depression-era consumers could afford. Al knew that if Bemis was to survive, he had to find another product—one that was in demand and required the woodworking skills of his employees. He chose toilet seats.

Crocker Chair Company had been making seats for Kohler Co. toilets. It was a profitable account that Al sought. So he negotiated the purchase of Crocker's seat-making machinery, tools, dies and patents for $2,000. Al hoped to gain the Kohler account as well. Unfortunately, that business went to another area company.

Nevertheless, using Crocker's equipment and patents as a springboard, Bemis launched a new product line and entered a new industry—plumbing products. With that move, Bemis joined other companies that survived the Great Depression by capitalizing on their strengths and applying them to new ventures.

The Great Depression

- 9,000 banks failed
- 9 million savings accounts were wiped out
- The price of stock fell 40 percent
- 86,000 businesses failed
- Wages declined by an average of 60 percent
- The unemployment rate went from 9 percent to 25 percent
- 15 million Americans were without jobs

A company catalog, circa 1933, explained that Bemis purchased machinery, patents and copyrights from Crocker Chair Company: "We feel confident that we are in a position to manufacture toilet seats to meet the requirements of the plumbing trade."

A. Carl Jensen

Carl Jensen joins the company

Al Bemis took a risk when his firm began producing toilet seats in 1932. Bemis Manufacturing Company was skilled at woodworking but had no experience in the plumbing products industry. The company also had no knowledge of sales channels and lacked the safety net of a major account.

At that time there were about 30 seat manufacturers in the United States, and Al had to find a way to differentiate his product from the competition.

While he had equipment and patents from Crocker Chair Company and its design for a unique wood interlocked toilet seat that resisted splitting, Al did not have the Crocker customer list. A man named A. Carl Jensen did.

A former purchasing agent for Crocker, Carl was without a job when the company declared bankruptcy. But he retained his industry connections, expertise and Crocker's customer list. With those assets, Carl pitched his services to area furniture manufacturers. No one was interested—until Carl met with Al Bemis.

Al knew that marketing Crocker's interlocked seats to the company's former customer base might give Bemis a sales advantage. So he entered into a partnership with Carl in 1932, granting Carl one-quarter ownership of the Bemis business and naming him executive vice president. Carl was appointed treasurer and was charged with leading the company's sales function.

Seats gain a foothold

The addition of Carl Jensen and the Crocker customer list bode well for Bemis and helped establish the company in the plumbing products industry. Depression-era federal legislation also gave Bemis and its fledgling toilet seat product line a much-needed boost.

The National Recovery Act, or NRA, was enacted in 1933 to encourage manufacturers of similar products to work together to maintain fair pricing in the face of plummeting business opportunities. The NRA helped

Oleida almost thwarts toilet seats at Bemis

Former Crocker purchasing agent A. Carl Jensen had the knowledge Al Bemis needed to make the toilet seat business a success. But a partnership between the two was nearly thwarted by Al's wife Oleida. This is the story of how Carl came to Bemis, as told in recent years by retiree John C. Kalk, a 32-year employee of Bemis:

Oleida Bemis

"Carl said to Al Bemis: 'I would like to get 25 percent (of the business). I would like to get in on the business and make my efforts worthwhile.' Al Bemis said it looked like a good idea, but his wife Oleida said: 'No way are we going to give you any part of the Bemis Company. It's our company, and Al isn't going to give you anything.' "

As the story goes, Al told his wife that the company was in danger of closing its doors without Carl's knowledge and industry contacts. So Oleida reconsidered, and she and Al agreed to grant Carl one-quarter ownership in the family business and pay him a salary of $100 per month.

1929	1930	1932	1933	1935	1935	1938	1938
The stock market collapse on "Black Tuesday" signals the onset of the Great Depression.	America toasts to 10 years of Prohibition. The law remains for three more years.	Bemis purchases toilet seat machinery and patents from Crocker Chair Company.	Franklin Delano Roosevelt is elected to first term as U.S. President.	Bemis wins the account to make seats for Kohler Co. toilets.	F.K. "Pete" Bemis becomes production manager at Bemis.	The BWA, or Bemis Workers Association, is founded.	The Fair Labor Standards Act is signed into law, setting the minimum wage at 25 cents an hour.

restore stability to the plumbing products industry. That stability enabled Bemis to flourish and gain significant market share—including the much-coveted Kohler business.

Kohler had grappled for several years with toilet seat suppliers that were struggling during the Depression. The company sought a stable supplier, one that had the expertise and equipment to manufacture quality seats for Kohler toilets. That company was Bemis.

Carl Jensen was instrumental in Bemis gaining the Kohler account in 1935. So was the company's pricing strategy that underbid competitors. But perhaps the strongest factor in winning the business was the Bemis newly gained expertise in making toilet seats to demanding Kohler standards.

With this and other sales gains, Bemis soon recognized that its future was not in coaster wagons or novelty furniture—but rather in toilet seats.

Management at Bemis, circa 1933, included company President Al Bemis (center) and Executive Vice President/Treasurer A. Carl Jensen (right). A few years later, Al's only son "Pete" (left) would graduate from law school and join the family business.

Second generation emerges

In 1935 Al Bemis' son, Pete, graduated from the University of Wisconsin Law School. The country was still in the throes of the Great Depression—a tough time to start a law practice. In later years Pete would recall his memories of two other struggling college graduates he was acquainted with: "Only one got a job and the other was a soda jerk."

Al helped his son launch a career with the family business instead of a law practice. He convinced Bemis co-owner Carl Jensen to disregard the company's non-nepotism agreement. Within months of his graduation, Pete came to Bemis as production manager.

F.K. "Pete" Bemis

His first task was to evaluate the plant's manufacturing processes. Pete found a factory and staff divided between crafting the company's old novelty furniture line and its new more lucrative toilet seats.

Touring the plant, Pete watched frustrated workers break 80 pieces of glass as they struggled to make a single coffee table. This was enough to convince Pete that the company should direct its focus to toilet seats.

So, in a bold move, Pete instructed workers to feed wood remnants of the furniture business into plant boilers. This act signaled the demise of what Bemis Manufacturing Company had long been—a furniture factory. It also marked the birth of the company's new identity as a plumbing products supplier.

Bemis and Kohler—partners in prosperity

When Bemis Manufacturing Company secured the account to make seats for Kohler Co. toilets in 1935, it was the beginning of a long-standing relationship between the two companies that continues today. Kohler has the distinction of being the longest continuous Bemis customer and its first major account. Kohler also helped Bemis strengthen its position in the plumbing products industry. In the midst of the Great Depression, the Kohler account meant something more tangible and highly valued—jobs and income for Bemis employees who otherwise might have struggled to make a living.

Cash on the barrelhead

During the lean Depression years, businesses struggled to remain open and went to great lengths to meet their debts when cash was short. Bemis was no exception.

Company management turned to unconventional methods to pay employee wages and purchase raw materials and equipment. Those included a loan from a local tavern owner.

During the Depression Bemis employees donated $1 apiece for the company's annual picnic. Shown above in this 1930s photo are Bemis Vice President/Treasurer A. Carl Jensen (left) and fellow workers enjoying a company outing.

One evening in 1935 a distressed Carl Jensen stopped at a Sheboygan bar and confided to owner Leo Zorko that he had issued the company payroll without sufficient cash reserves. As treasurer of Bemis, Carl feared he might go to jail for writing bad paychecks.

In a generous move Zorko led Carl to his upstairs apartment, where he pulled back the living room carpet to reveal cash spread on the floorboards. It was a safer place to store savings, Zorko mused, than in banks which were failing by the score. Zorko gave Carl cash to cover the payroll and asked him to repay it when Bemis was in better straits. It took only a handshake and no collateral to seal the agreement between Zorko and Carl.

The Great Depression changed more than banking habits. It also altered the way business was conducted.

Extending credit was risky, so "cash on the barrelhead" became the norm. Manufacturers including Bemis even bartered to obtain much-needed supplies.

Bemis acquired lumber for toilet seats from timber mills in northern Wisconsin. It was a desolate area, rich in lumber, yet lacking in the culinary basics common to Sheboygan. Instead of paying cash for lumber, a resourceful Carl would offer sausage and cheese as partial payment to lumber companies and their hungry employees.

Wood—the cornerstone of Bemis

At the foundation of Bemis advances in product design and manufacturing is a simple raw material—wood.

Wood has been a cornerstone of Bemis since its early days at the start of the twentieth century as White Wagon Works. It was then that the company constructed children's wagon bodies of wood.

Dependence on pulp and wood products continued in the 1910s and 1920s as Bemis-Riddell Fibre Co. manufactured popular fibre furniture woven from treated paper.

Even the novelty furniture that was part of the early Bemis product offering was made of wood, as were the first toilet seats that later became the company's focus—and remain its most enduring product.

Bemis at the end of the Great Depression—In the mid- to late-1930s, Bemis chose to focus on toilet seats and cease manufacturing novelty furniture. The company retained hampers, benches and stools as part of its product line. Pictured here are management and employees of that era including Al Bemis (left), Pat Babb (third from left), A. Carl Jensen (center with hat), F.K. "Pete" Bemis (behind Carl's left shoulder), Walter Trowbridge (at Carl's left), and Alvin Popp (far right).

Disaster strikes

Throughout the mid- to late-1930s Bemis continued to manufacture and market toilet seats. The U.S. economy was slowly improving, and business at Bemis was steady.

But the company's calm was shattered by a tragic accident in 1938. The day after Thanksgiving, a boiler room explosion shook the Bemis plant. Four men were inside when the blast occurred without warning, including company Secretary and Plant Superintendent Walt Trowbridge.

Walt's clothes ignited when flames shot out of the boiler. In panic he jumped out a window and landed in the frigid Sheboygan River just outside the plant. His back, face and hands were severely burned, yet the river saved Walt's life.

Of the other men caught in the explosion, one perished and two others were injured.

Despite the severity of the incident, the Bemis factory sustained minor damage, as the blaze was extinguished in short order by the Sheboygan Falls Fire Department, located just a block away.

Bemis recovered quickly from the blast and looked to a more hopeful future. In 1938 management realigned the organization, in part to terminate the old Wagon Works name. This included dissolving the Bemis Manufacturing corporate entity first founded in 1925 as a sales organization for White Coaster Wagon Works. All BMC assets were transferred to White Coaster Wagon Works. The Wagon Works' name was then officially changed to Bemis Manufacturing Company, thus creating the business entity that exists today.

Sheboygan Falls Blast Kills 1 And Injures 3

(Cintinued from page 1)

river. Both men are reported doing fairly well today.

The flaming blast that shook the building also set it afire. The Sheboygan Falls fire department was called at once and succeeded in extinguishing the flames before a great deal of damage was done by fire.

Factory officials today puzzled over the cause of the explosion. No one seemed to know just how it happened. Some recalled instances in other cities where dust-filled air had been ignited and caused similar blasts.

Benjamin Goll lived with his family for the last 15 years at 512 Oak st... Falls. For ... he worked a... factory...

May 29, 1884, the son of ... and Mrs. Fred Goll. In ... he married Esther Pear... the couple lived at ... until coming to Sh... Falls 15 years ago.

He is survived by one son, Fred, of ... Falls; one grandso... ters, Mrs. Herm... Mrs. Henry Klei... August Kramer, ... Mrs. John Stra... Scott, and Mrs. ... kin, Cascade; ... John Goll, of ...

Funeral se... ducted at 1... at the Wal... eral home ... The Rev. ... the Sheb... church ... will be ... tery.

Walt Trowbridge was seriously injured in 1938 when dust ignited and exploded in the boiler at the Bemis Broadway Avenue plant (later known as Plant A) in Sheboygan Falls. After he recovered, Walt resumed his responsibilities as plant superintendent and secretary. A newspaper clipping from The Sheboygan Press, *November 26, 1938, details the explosion.*

Emblazoned with the name "Bemis Mfg. Co.," a new smokestack was erected on the roof of Plant A in Sheboygan Falls in 1948. This building was the former home of Bemis predecessor White Wagon Works. Shown here is Bemis Secretary Walt Trowbridge (left) with an unidentified worker.

CHAPTER 3

Bemis and World War II

The end of the 1930s brought an era of hope as America began its recovery from the Great Depression years. That spirit was mirrored at the grand and futuristic 1939 New York World's Fair, where "The World of Tomorrow" displayed the latest technological wonders, and a physicist named Albert Einstein demonstrated how cosmic rays worked.

Bemis Manufacturing Company was hopeful too. The company had struggled to stay in business during the country's worst economic period in history—a time when nearly 100,000 businesses closed, thousands of banks failed and millions were unemployed.

But it was the dawn of a new era, one of anticipation and the promise of success. And Bemis looked to make the most of that time.

Threat of war imminent

Despite this hopeful sentiment, there was a cloud on the nation's horizon. War broke out in Europe in 1939, and Americans kept a wary eye on that conflict, speculating what might happen if the country entered World War II.

In those post-Depression years, the United States supported the European war from a distance. The nation's industrial productivity soared, and many factories were transformed into high-intensity arms-production plants. Scores of the unemployed returned to work to construct FDR's "arsenal of democracy" for Great Britain and the Allied Forces in Europe.

Supporting the war effort

"Use it up, wear it out, make it do, or do without" was the rallying cry of the time, encouraging people to contribute to the war effort.

- Each American was allotted three pairs of leather shoes per year
- Kitchen fat was collected and processed for explosives
- Everything from old tires to boots was donated at scrap rubber drives nationwide
- People brought paperbacks to victory book rallies across the country
- Sales of ladies' trousers skyrocketed as women went to work in factories
- Shirley Temple, Ingrid Bergman and other Hollywood stars urged moviegoers to buy war bonds

Bemis employee Virginia Bonnet, circa 1948, sands a toilet seat ring coated with primer at the company's Broadway Avenue plant in Sheboygan Falls. After sanding, an enamel paint topcoat was applied to finish the seat.

Yet not all manufacturing efforts were focused on military buildup. Common and indispensable products including apparel, appliances and automobiles continued to be produced, and Bemis continued to manufacture what had become its flagship product—toilet seats.

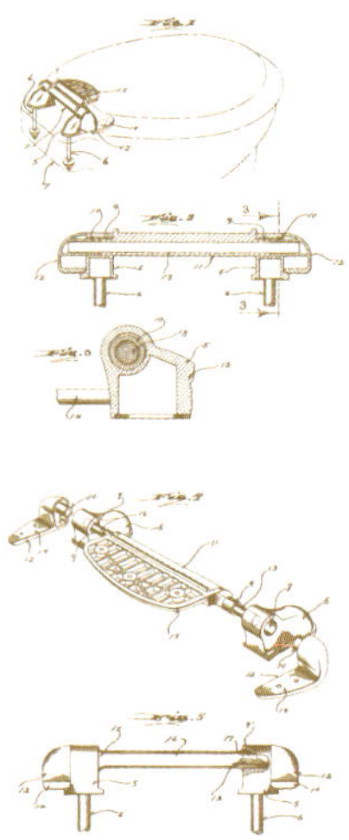

Inventor and Bemis business partner Harold Watson submitted these sketches of a plastic toilet seat hinge along with his application for a patent in 1941. The patent was granted, and Harold's design became the first successful plastic toilet seat hinge used at Bemis.

Toilet seats at risk

World War II threatened more than world democracy, the nation's security and American lives. It threatened seats for toilets.

In the 1940s, toilet seats—including the scores that Bemis made each year—were constructed of wood. Rings and lids were joined with hinges made of brass. But brass was fast becoming scarce with much of it diverted to manufacturing shell casings and other defense-related products.

Company President Albert Bemis recognized that if the nation were to enter the war in Europe, toilet seat manufacturing at Bemis would be in serious jeopardy without brass for hinges. This potential critical shortage left Bemis Manufacturing vulnerable, and Al knew he had to find a substitute hinge material if his company was to survive.

A little luck and a new hinge

Little did Al suspect that a chance luncheon with a business associate would also serve up a viable alternative to brass hinges. More important, it would help Bemis leap beyond the competition to take a leadership role in the plumbing products industry.

The answer was in a new petroleum-based material—one that was versatile, plentiful and inexpensive. It was known as plastic.

Al's lunch acquaintance, a district agent for General Motors, was familiar with plastic. But the agent, Harold Watson, offered Al something more vital than his passing knowledge of this new material. He proposed a potential manufacturing source for plastic hinges.

Encouraged by Harold's knowledge, business contacts and interest in crafting an alternative hinge for Bemis, the company financed a development project with Harold in charge. The project grew to what would later become Falls Plastics Company of Sheboygan Falls, Wisconsin.

Falls Plastics Company—the first Bemis subsidiary

Falls Plastics Company was founded in 1942 as the first Bemis subsidiary business—one that would provide the toilet seat manufacturer with much-needed plastic hinges during the World War II years.

Bemis used a plastic hinge design for just a brief period of time during the war, yet it was that hinge—and Falls Plastics' role in supplying it—that helped Bemis continue to manufacture toilet seats in the midst of a war-induced brass shortage.

Falls Plastics Company and its plastic hinge also enabled Bemis to leap ahead of the competition that was left struggling with how to make toilet seats without brass hinges during the war.

Once the war was over, Bemis abandoned the plastic hinge and returned to brass for a time until technological advances made a superior plastic hinge possible. But Falls Plastics Company and its president—plastic hinge inventor Harold Watson—are noteworthy for having played an important role in the history of Bemis Manufacturing Company and the prominent stature it enjoys today.

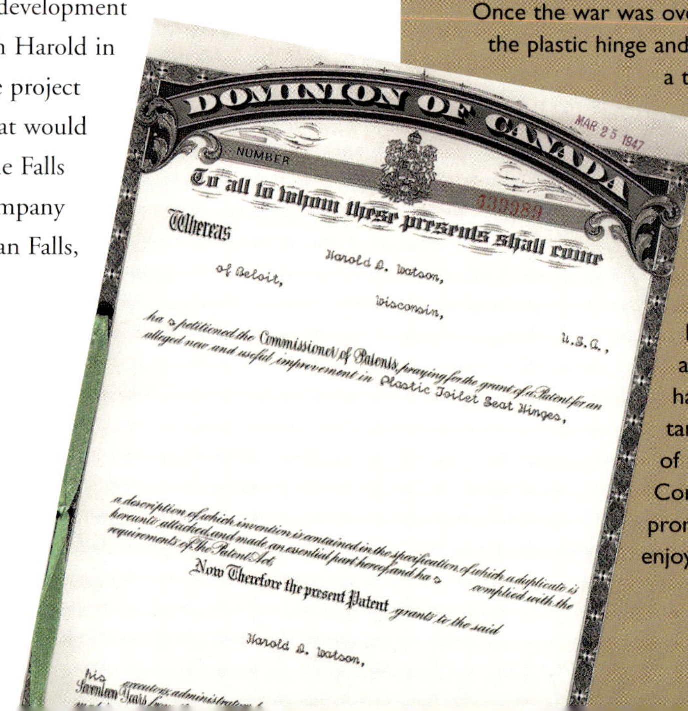

DOMINION OF CANADA

MAR 25 1947

NUMBER 339989

To all to whom these presents shall come

Whereas Harold D. Watson, of Beloit, Wisconsin, U.S.A.,

has petitioned the Commissioner of Patents praying for the grant of a Patent for an alleged new and useful improvement in Plastic Toilet Seat Hinges,

a description of which invention is contained in the specification of which a duplicate is hereunto attached and made an essential part hereof, and has complied with the requirements of the Patent Act,

Now Therefore the present Patent grants to the said Harold D. Watson,

Harold Watson to the rescue

Harold Watson, an innovative man and budding inventor, worked with an Illinois molder to develop a viable plastic hinge for Bemis. The project spanned months of trial and error with a variety of plastics. Initial efforts created a unique one-piece hinge of Bakelite, a thermoset phenolic polymer. In testing the hinge would not support the weight of a 250-pound man. Its excessive price tag also made the Bakelite hinge project a dismal failure.

Harold tested hinges made of other plastics. While none had the superior performance of brass, he settled on one that functioned adequately. Recognizing that the nation was on the cusp of entering the European war, Bemis ordered 250,000 of these plastic hinges at a price of 36 cents apiece.

A leap beyond the competition

Harold Watson filed for a patent on his plastic toilet seat hinge design in December 1941. It was just in time to help preserve the Bemis toilet seat business, as just days earlier the United States Congress declared war on Japan in the wake of the surprise attack on Pearl Harbor.

With the nation now at war, toilet seat manufacturers were left scrambling to find viable alternatives to brass hinges. But Bemis had a new plastic hinge in plentiful supply and a growing customer base that they could continue to serve. These advantages helped the company survive in the face of stiff competition, gain market share, and expand its operations and influence in the nation's plumbing products industry.

Workers from the Bemis Hinging Department take a break outside the Broadway Avenue plant in Sheboygan Falls, circa 1948.

Bemis sends its own to war

Just seven months after the United States entered World War II, Pete Bemis received his draft notice. It was June 1942, and Pete was 32 years old, treasurer of Bemis and father of two toddlers. Like so many men in those war years, Pete left his comfortable life in Sheboygan and joined 16 million Americans in uniform who were fighting for the world's freedom.

Fortunately, Pete received a domestic assignment, in part, because of his law degree. For a portion of his enlistment, Pete was

1939	1941	1941	1942	1945	1946	1947	1948
WWII begins after the Nazis invade Poland.	BMC finances Falls Plastics Company to make plastic toilet seat hinges.	Hollywood hits *Casablanca* and *The Road to Morocco* entertain Americans adrift in a world at war.	F.K. "Pete" Bemis is drafted into the U.S. Army.	FDR dies and is succeeded by Harry Truman as U.S. President.	Albert Bemis dies at age 68, and his son Pete Bemis becomes company president.	Jackie Robinson breaks the color barrier as the first African American to play major league baseball.	Bemis expands to former farmland west of Sheboygan Falls and names the new complex Plant B.

Family man

When Pete Bemis was drafted into the U.S. Army in June 1942, he had been married for four years to Grace. The couple had two children, Patty, 3, and Dick, 1. Pete was home on frequent leave during his enlistment and even moved his family with him to Washington, D.C., when he was stationed there with the Pentagon's Department of Procurement, Price Analysis Branch.

assigned to the Pentagon in Washington, D.C., where his days were spent reviewing purchase agreements for materials needed to support the war effort.

During Pete's three-and-a-half-year wartime service, his position as treasurer of Bemis Manufacturing was assumed by company Vice President Carl Jensen.

Life during wartime

Employees at Bemis Manufacturing Company kept a watchful eye on the war as one of their own, company heir Pete Bemis, served in the Army. But business carried on in Pete's absence.

F.K. "Pete" Bemis, circa 1942

Bemis continued to gain market share and expand manufacturing capacity to meet its growing customer base. Those customers now included the United States military.

A necessity in peace or wartime, toilet seats were a commodity in demand by U.S. troops serving in makeshift living quarters across the globe.

To meet growing military needs, Bemis—like many manufacturers of the day—refocused a portion of its production to war-related products. These included toilet seats for army camps.

Bemis also produced folding chairs for easy transport and use at portable military living stations worldwide. Skilled company woodworkers shaped chair component parts, then shipped them to another area firm, Richardson Industries, for staining and assembly. Bemis crafted thousands of these chairs to support the war effort. It was a return to the company's furniture-making roots as Bemis-Riddell Fibre Co. of two decades earlier and as Bemis Bros. & Crocker Chair Company in the late 1800s.

An era vanishes

Three months after World War II ended in September 1945, Pete Bemis was discharged from the Army—just in time to celebrate Christmas at home in Sheboygan.

His daughter Patty was now 6, and son Dick was 4 years old. A third child, Catharine, who had been born while Pete was in the service, was 2 years old. Eventually the family would grow to five children, with son Peter coming along in 1947, and Mary Louise, or LuLu, joining the family in 1949.

The season of tranquility that Christmas 1945 brought was profound and far-reaching. The world was finally at peace. Democracy had been preserved, and American troops, including Pete, had returned home.

Yet there was sadness in the Bemis family, which had lost matriarch Oleida two years earlier at age 65. Oleida had been instrumental in the founding and early management of Bemis Manufacturing Company.

Her husband Al lived just three years after Oleida died. Still president and chairman of Bemis, he passed away suddenly at age 68.

Two months earlier, Al's older sister Luella Bemis Trowbridge—company accountant and a founding partner in White Wagon Works—had also passed on.

Sadly it was the end of an era for Bemis Manufacturing Company—a transition from its founders to the next generation. That second generation would take Bemis to accomplishments not yet imagined.

Three generations gather for a rare photo, circa 1945. F.K. "Pete" Bemis in uniform during his Army service in World War II is joined by young son Dick and father Al.

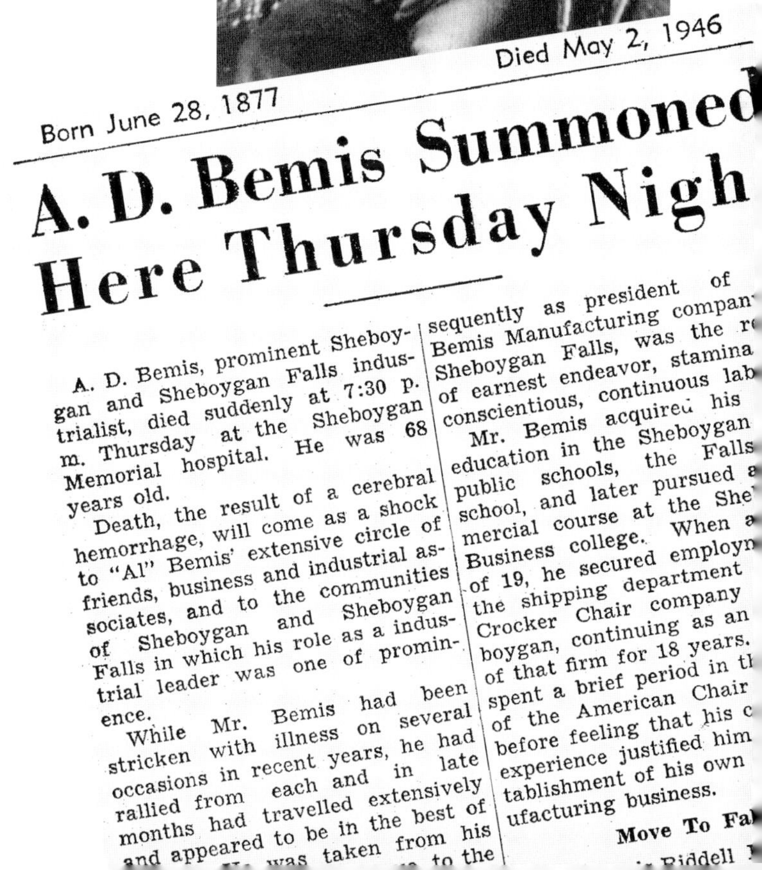

Died May 2, 1946

Born June 28, 1877

A. D. Bemis Summoned Here Thursday Nigh

A. D. Bemis, prominent Sheboygan and Sheboygan Falls industrialist, died suddenly at 7:30 p. m. Thursday at the Sheboygan Memorial hospital. He was 68 years old.

Death, the result of a cerebral hemorrhage, will come as a shock to "Al" Bemis' extensive circle of friends, business and industrial associates, and to the communities of Sheboygan and Sheboygan Falls in which his role as a industrial leader was one of prominence.

While Mr. Bemis had been stricken with illness on several occasions in recent years, he had rallied from each and in late months had travelled extensively and appeared to be in the best of ... was taken from his ... to the

sequently as president of Bemis Manufacturing compan Sheboygan Falls, was the r of earnest endeavor, stamina conscientious, continuous lab

Mr. Bemis acquired his education in the Sheboygan public schools, the Falls school, and later pursued a mercial course at the She Business college. When a of 19, he secured employn the shipping department Crocker Chair company boygan, continuing as an of that firm for 18 years. spent a brief period in tl of the American Chair before feeling that his c experience justified him tablishment of his own ufacturing business.

Move To Fa

Riddell

In Albert Bemis' obituary, which appeared on May 3, 1946, The Sheboygan Press *credited his steady rise in the business world to "earnest endeavor, stamina and conscientious, continuous labor."*

F.K. "Pete" Bemis assumed the role as company president in 1946 after the sudden death of his father Al.

Expansion to Plant B

Upon Al Bemis' death in May 1946, son Pete, back from the Army just six months, assumed his father's duties as president and chairman of Bemis Manufacturing Company. Together with company Vice President Carl Jensen and Secretary Walt Trowbridge, Pete continued to build the business, which included toilet seats and other wood plumbing products such as toilet tank covers.

The company's manufacturing and management operations were still situated at its original location, the old Broadway Avenue plant in Sheboygan Falls, which had been home to the former White Wagon Works. But the time had come to expand beyond this cramped location.

That expansion would be to farmland west of downtown Sheboygan Falls on property owned by a local banker named John Heronymus.

A barn, two silos and a chicken coop

John Heronymus used his property on Mill Street in Sheboygan Falls to raise chickens, plant victory gardens in the spirit of postwar America and entertain clients.

In 1948 Pete Bemis negotiated the sale of the entire Heronymus parcel, which had two grain silos, a barn, a chicken coop and a farmhouse. This land would later become the site of Bemis Plant B, with the old Broadway property being renamed Plant A.

Rather than demolish the farm buildings, Bemis put them to good use—at least until the company could afford to replace them with more suitable structures. The barn was converted to a manufacturing site where Bemis workers coated toilet seats with a mother-of-pearl finish using a highly flammable process. A door to an outdoor hayloft offered an exit just in case the process sparked a blaze. Fortunately, workers never had to use this makeshift fire escape.

The new Plant B site produced more than toilet seats in its early days. It yielded freshly laid chicken eggs. The company had chosen to keep poultry in the Heronymus chicken coop for a time, and Pat Babb, foreman of the Bemis wood shop, would gather eggs to distribute to fellow employees.

Today that chicken coop is the site of Bemis corporate headquarters—a facility that manages the company's 3,000 employees and multiple business units serving markets across the globe.

In 1942 Bemis purchased an old pea-canning factory (right) and its railroad right-of-way for rail lumber deliveries. Several years later the company bought an adjacent farm, and the area became known as Plant B. This view is looking north from the intersection of Mill Street and Sheboygan County Trunk Road PP. It shows storage silos and stacks of lumber for toilet seats. This lumber air-dried for a time before it was transported to Plant A for kiln drying and processing.

Bemis worker Roger O'Reilly carefully inspects a toilet seat at the company's Plant B, circa 1960. Seats were compression molded from wood flour and resin. These new seats and the technology used to manufacture them positioned Bemis for success in the plumbing products industry.

CHAPTER 4

The age of plastics arrives at Bemis

The 1950s were a time of tranquility and unprecedented prosperity as the United States secured new world leadership in technology, manufacturing and the arts. Elvis rocked America with risqué pelvic gyrations, and Marilyn made moviegoers swoon. Two-tone convertibles with whitewall tires were the rage of the road. Television entered middle-class homes, and the Soviet satellite Sputnik sparked a science fiction frenzy.

A postwar baby boom exploded as young World War II veterans married and settled in sprawling new suburban communities. As housing and construction soared, companies like Bemis that supplied building-related products prospered.

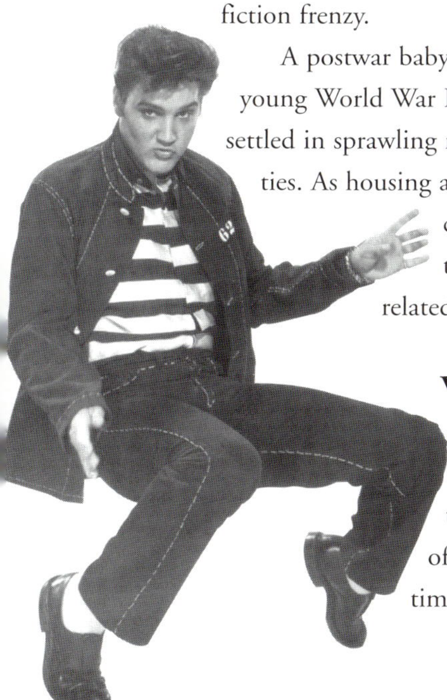

Waste not, want not

In the early 1950s Bemis manufactured toilet seats exclusively of laminated birch. The time-intensive process included drying rough lumber, sawing boards into pieces, gluing them together with dowels imbedded for strength, shaping seat components, then sanding and painting to finish the product. This method was laborious and wasteful, and that wood waste represented lost dollars to Bemis.

So it was with great interest that company management considered a newly emerging method for manufacturing seats. This novel process blended sawdust with plastic resin and cured the mixture under intense heat and pressure. The resin bonded with the sawdust to create a product that was denser and heavier than wood. The process was known as compression molding.

A future in compression molding

Company President F.K. "Pete" Bemis recognized the advantages of converting Bemis toilet seat operations from woodworking to compression molding.

The new technology made more efficient use of raw materials and created a more durable seat because compression molded products were moisture-resistant and wouldn't split apart like those made of birch.

Compression molded toilet seats also

"Be Seated by Bemis" was a slogan that appeared on the company's delivery trucks for decades. While its origin is undocumented in Bemis historical records, the slogan expresses the company's focus on toilet seat manufacturing for the plumbing products industry.

Engineers Walt Trowbridge (top) and Jack Nelson were vital in transforming Bemis from a woodworking company to a molder of plastics and plastic/wood composites.

offered greater design flexibility than laminated wood, and those enhanced design options could prove a competitive advantage for Bemis.

Pulverizing lumber into fine sawdust, or wood flour, also meant Bemis was less vulnerable to price fluctuations of northern hardwood. Other toilet seat manufacturers had begun converting their operations to compression molding, and Pete knew that to remain competitive, Bemis would have to do the same.

Trial and error

Developing the optimum mixture of plastic resin and wood flour would be a challenge. But it was a research project that the Bemis senior staff undertook with fervor in the early 1950s.

For months Pete, Walt Trowbridge and a new hire, Pete's brother-in-law Jack Nelson, searched to find the best formulation for a compression molded toilet seat, one that when cured would offer the utmost strength and durability.

Together the novice plastics formulators set up their experimental lab in the basement of Plant A on Broadway Avenue in Sheboygan Falls. They used unconventional equipment to perfect the wood flour and resin recipe.

To provide heat for curing, Pete, Walt and Jack purchased a secondhand kitchen stove from Ebbers Hardware Store in Sheboygan Falls. They crafted a mold from cast iron and used an old cement mixer to blend varying proportions of wood flour, resin and additives.

The team collaborated with Sheboygan phenolic resin supplier Plastics Engineering Company. In time they came upon a recipe and a curing method that worked, and coined the new process "Molcor."

The Bemis management team of the 1950s included (from left) Walt Trowbridge, F.K. "Pete" Bemis and Carl Jensen.

1951	1951	1951	1954	1956	1959	1960	1961
I Love Lucy and *The Roy Rogers Show* debut on network television.	Bemis begins experimenting with a new compression molding process for making toilet seats.	Sales and engineering departments are expanded to increase the toilet seat product line.	A new vaccine for polio is unveiled in hopes of eradicating the disease most feared by parents.	Elvis Presley explodes onto the national charts with *Heartbreak Hotel, Don't Be Cruel* and *Blue Suede Shoes.*	Bemis is awarded its first bowling pin contract from Brunswick.	Bemis purchases its first injection molding press—a used Lester machine.	Astronaut Alan B. Shepard Jr., America's first man in space, proclaims "What a beautiful view!"

Engineering—a key to Bemis success

New products and processes have sustained Bemis and helped the company grow and flourish during its 100-year history. Pivotal to this has been the company's topnotch engineering staff, the first of which was Walt Trowbridge, Bemis secretary and a civil engineer by trade. Other early engineers were Jack Nelson and Jeff Trowbridge.

Jack, a brother-in-law of F.K. "Pete" Bemis, came to the company in September 1952. Jeff, son of Walt Trowbridge, joined Bemis a few months later. Together, Walt, Jack and Jeff developed and perfected manufacturing processes that made compression and injection molding a reality for the company.

Because of their vital contributions, Bemis was able to enter new arenas in the 1950s and 1960s, make the most of market opportunities and re-engineer itself from a woodworking company to a molder of plastics and plastic/wood composites.

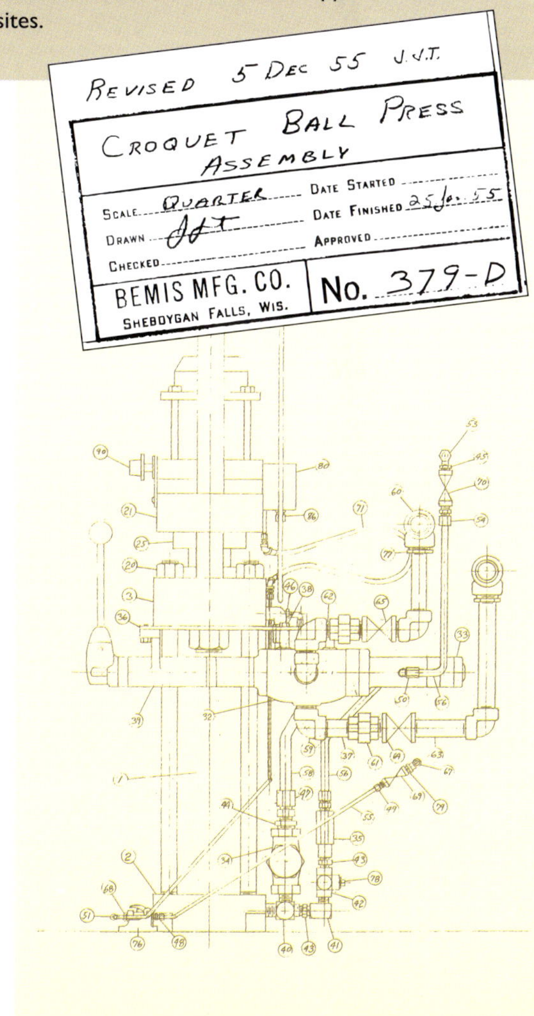

Molcor comes to Bemis

By 1953 Bemis was manufacturing all toilet seat rings using the new Molcor compression molding process. The company continued crafting seat lids from laminated wood, since this piece was easier to shape than the donut-shaped ring portion.

Melding the Molcor seat to the laminated wood lid proved a success for Bemis. The company designed and built custom presses and molds to engineer Jack Nelson's specifications. Operators mixed components according to early trial-and-error formulations to produce 1,500 Molcor seats per day. Much of this original process—established nearly a half century ago—is still used at Bemis today.

The art of fine finishing

Bemis met with another challenge when operators attempted to apply a flawless enamel paint finish to the compression molded/laminated wood toilet seats. They hand-dipped seats into vats of paint twice for a two-layer base coat, then applied two spray top coats.

Jack Nelson recognized this four-layer finishing process was time-consuming, expensive and far from flawless.

So, together with newly hired engineer Jeff Trowbridge, Jack devised an automated flow coat procedure to apply primer and paint.

Operators suspended seats from hooks on a conveyor line, which moved through a

Trowbridge family legacy

Other than Al Bemis and his descendants, perhaps no family has been as instrumental in the growth and success of Bemis Manufacturing Company than the Trowbridges—Luella and husband Otis, their son Walter, and his son Jeff.

The Trowbridge family relationship with Bemis began in 1903, when Al Bemis' sister Luella and her spouse Otis became shareholders in White Wagon Works, the precursor to Bemis Manufacturing Company.

That business relationship flourished into the 1920s when Luella engaged the interest of Al in partial ownership of the Wagon Works, then worked with him to establish Bemis Manufacturing Company as its sales arm.

Luella's only child Walt was also active in company interests as secretary/treasurer and co-owner upon his mother's death in 1946. Walt was the first engineer at Bemis—and a critical contributor to the company's expertise in toilet seat manufacturing. He was also instrumental in its later transformation from a woodworking enterprise to a plastics molder in the 1950s and 1960s.

Badly burned in a 1938 boiler room explosion at Bemis, Walt nonetheless continued to commit himself to the company's interests until his death in 1962. He was 63 years old at the time and had spent 41 years at Bemis.

Walt's son Jeff, who joined the company in 1952, has also left a legacy at Bemis. A degreed mechanical engineer, Jeff designed the Bemis hinging line operation, plus the manufacturing processes for croquet balls and baseball bats. In 1997 Jeff celebrated 45 years at Bemis. He continued to work in engineering for the company well into his 70s.

While the Trowbridge family sold their ownership interest in Bemis during the mid-1980s, they continued to shape the company's future with nearly 100 years of vital contributions.

Three generations of the Trowbridge family include (from top) Otis, circa 1890s, his son Walter, circa 1920s, and Walt's son Jeff, circa 1950s.

series of paint tunnels. There the seats were flow coated with primer and undercoat. Operators then applied a spray top coat by hand.

This first flow coat production line was constructed in the basement of Plant A. Building the line was an immense undertaking for Bemis, one that was expensive, risky and based on the premise that Jack and Jeff's prototype process could be effectively transferred to a high-speed production line. Company management found that it could.

Bemis used its novel flow coat technology to produce toilet seats with higher gloss and a smoother finish than any competitor. This paint finish was tops in the industry and set Bemis apart from other toilet seat manufacturers.

Contract Division established

Molcor toilet seats and compression molding proved such a success that by the mid-1950s, Bemis chose to make seat lids as well as rings from the wood flour and resin mix. Yet the company still employed scores of woodworkers and had equipment standing idle. These included rough lines, dimensional lines and a gluing operation.

So Bemis scrambled to put these assets to use by launching a Contract operation to supply custom wood components to manufacturers in other industries. These included steering bars and slats for children's sleds, plus tablet arms, seats and desktops for school desks.

This business lasted just a few short years in the mid-1950s until plywood offered a cheaper alternative to laminated hardwood. With its woodworking business under threat from plywood,

Steering bars for children's sleds made by Garton Toy Company of Sheboygan (opposite page) were one of the first Bemis Contract products produced. Croquet and boccie balls, plus beer taps (opposite page below), soon followed—but unlike solid wood sled components, these were made using the new Bemis Molcor compression molding process.

A 1950s trade show appearance by Bemis *shows the company's diverse products—Bemis-branded toilet seats and a host of Contract items including billiard balls, bowling pins, and croquet mallets and balls sold to original equipment manufacturers nationwide.*

Bemis opted to expand its sales efforts to reach nationally known manufacturers of toys, housewares, sports equipment, furniture and appliances. These potential accounts required a skilled salesman. The company chose Gerald (Jere) Sampson.

Known around Bemis as "Mr. Bowling Pin," salesman Gerald (Jere) Sampson nailed the company's first order for bowling pins in 1959. Jere helped build sales of other Bemis wood products over four decades until he retired in 1989.

A social acquaintance of Pete Bemis and Carl Jensen, Jere came to Bemis in 1957 and set his sights on expanding the Contract business. He succeeded in a big way by obtaining an initial order from Brunswick for one million bowling pins.

Other all-wood products fabricated by Bemis and sold by Jere included croquet balls, hickory baseball bats, billiard balls and bowling pins. These toys and sporting goods formed a solid secondary business operation for Bemis, with the company's primary focus remaining on toilet seats for the plumbing products industry.

The two Bemis businesses—woodworking and compression molding—supported one another and gave the company stability in a rapidly changing marketplace.

New toilet seat brand, new product line, new sales

By the mid-1950s Bemis management recognized the company could broaden sales beyond plumbing wholesale distribution to a new channel—retail.

Department, hardware and specialty chain stores carried toilet seats that were purchased by homeowners as replacements to well-worn household seats. But Bemis sales reps were accustomed to calling on plumbing wholesalers and not housewares buyers.

To sell seats to retail stores, Bemis would have to create a new product line with a new name that wouldn't conflict with the Bemis-branded wholesale business that the company had worked decades to build.

Sales Manager John Kalk was charged with finding a suitable name for the new retail line, one that would appeal to a female shopper with a different mindset than the plumbing wholesaler—a brand that would denote high style and top quality. He chose Mayfair after the posh London suburb.

John worked with Pete Bemis and Carl Jensen to create the new Mayfair product line, initially comprised of a few seat styles in a few colors. The line was featured in product literature and marketed to housewares buyers throughout the Midwest in the latter half of the 1950s.

For Bemis the launch of Mayfair signaled the birth of one of the bath industry's most well-known brands, one that has become a dominant name in toilet seats for today's retail bath channel.

Solid plastic seats enter the picture

While Bemis was perfecting its compression molding operations in the 1950s, a competitor's solid plastic product captured the attention of the industry.

Oscar Olson had been molding all-plastic toilet seats for nearly a decade and had a monopoly on this durable, water-resistant alternative. As the price of plastic dropped, Olson's seat garnered more sales for his company, Olsonite.

Pete Bemis knew that to remain competitive and offer a full product line of toilet seats, Bemis had to develop an all-plastic product that could go head-to-head with Olsonite. But converting Bemis toilet seat operations from laminated wood to Molcor had been a formidable task for Pete and his staff. Molding plastic seats would prove an even bigger challenge.

Staff engineers Jack Nelson and Jeff Trowbridge pioneered the process as they had when developing the Bemis Molcor seat. First attempts were to mold a plastic shell around a compression molded inner core. But the shell wouldn't adhere to the wood-based core.

Molding a solid plastic seat seemed to be the best option, but Bemis owned just one injection molding press in those days—a small, used Lester machine purchased in 1960 to mold hinges for Molcor seats.

So Bemis subcontracted with another Wisconsin molder, Milwaukee Plastics, which had the equipment and expertise to develop an all-plastic toilet seat. That first plastic seat was the Bemis Model 800, molded by Milwaukee Plastics from a die designed by Jack Nelson.

Bemis salesman John Kalk (left) discusses toilet seat features and benefits with his boss, Carl Jensen, circa 1955. Carl and company President Pete Bemis hired John in 1951 to boost toilet seat sales to both wholesale and retail distribution channels.

Bemis bowling pins strike success

Bemis woodworkers have crafted a host of products over the company's 100-year history. But perhaps none were as popular as bowling pins.

Bowling was a fast-growing sport in this prosperous postwar era, particularly with the advent of automatic pinsetter machines in the late 1950s. As the sport boomed, Bemis saw an opportunity to leverage its laminating expertise and grow with the bowling industry.

Skilled company woodworkers designed a maple laminated bowling pin comprised of 19 wood pieces. Sales staff then peddled the pin to Brunswick, the industry's largest user. Brunswick liked the Bemis pin—so much that they ordered one million of them!

Little did Bemis suspect that its bowling pin operation would support the company in the early 1960s. It was then that toilet seat competitor Olsonite slashed prices, making it tough for Bemis to compete. The company relied on profit from its bowling pin business to ride out this downturn in seat sales.

Bemis moved pin production to the north woods of Wisconsin in 1965 to be closer to the supply of hardwood maple and reduce costs. The company ceased pin production in the late 1970s. But while they were part of the product mix, bowling pins proved useful to Bemis and its growth through diversification.

A Bemis maple bowling pin block (left) joins a turned pin and a finished one sporting the Brunswick logo.

Future company leaders Peter and Dick Bemis *were youngsters in the 1950s when their dad F.K. "Pete" was building the family firm into a supplier of toilet seats and wood products for a host of industries. Shown here at Christmas 1952 are (from left) Peter and Dick, with sisters Patty, Catharine and Mary Louise, better known as LuLu.*

Injection molding signals new identity for Bemis

By the early 1960s products formerly made of wood were rapidly being re-engineered in plastic. Bemis products were no exception.

Plastic was versatile and plentiful. It was also less expensive than wood, both in raw material cost and labor requirements. Plastics represented the future, and F.K. "Pete" Bemis and his colleagues recognized this.

So in 1963 Bemis made a major commitment to enter the booming plastics industry. The company purchased a second, larger injection molding press—a used 1,000-ton Watson-Stillman machine at a cost of $15,000. With this new press Bemis moved all molding operations to its Sheboygan Falls Plant B and ceased working with Milwaukee Plastics.

As interest in plastic toilet seats grew, so did Bemis sales. Several building expansions to Plant B helped Bemis keep pace with growing demand as production of both plastic and compression molded seats soared.

The addition of injection molding was a noteworthy point in Bemis history. It broadened the Bemis toilet seat product line and its Contract Division capabilities. Now the company could offer its customers woodworking plus more cost-effective alternatives in plastic through two versatile manufacturing processes—injection and compression molding. More important, the transition to injection molding signaled the dawn of the new Bemis identity as a plastics molder.

An open toilet seat, complete with cover and hinges, graces early Bemis letterhead. This graphic and its prominent placement illustrates the company's emphasis on toilet seat manufacturing as central to its identity—an identity that has remained for most of its 100-year history.

This department store window display in Sheboygan, Wisconsin, shows many of the products that Bemis brought to market in the 1960s. They include plastic and compression molded toilet seats, toilet tank covers, toys like the Bemis Shut Box game, parts for the Playskool Tyke Bike, and sporting goods including croquet balls, Hickory Stick baseball bats, shuffleboard disks and bowling pins.

CHAPTER 5

A third generation enters

The 1960s were a time of tumultuous change, of freedom marches and racial riots, a conflict in Vietnam and a missile crisis in Cuba that nearly plunged the United States into a thermonuclear war with the Soviet Union.

It was an era of explosive emotion punctuated by presidential assassination, social injustice and materialism.

The turbulent times were reflected in the music of the age as a tousled songwriter named Bob Dylan revealed a powerful social conscience that stirred a generation. Among that generation were the two sons of F.K. "Pete" Bemis.

Dick and Peter Bemis were approaching adulthood in the early 1960s. Dick had graduated from Ohio's Denison University in June 1963—just months before President John F. Kennedy was assassinated in Dallas. Dick's brother Peter was looking forward to going off to school at Carroll College in Wisconsin. Both boys would come to play key roles in the futures of Bemis Manufacturing Company, Sheboygan Falls, and the hundreds of people who worked and resided there.

Dick Bemis

A job, a title and a future

Dick Bemis joined the family firm in 1963 on the sage advice of his father Pete. Single and having just earned a college degree in business administration, Dick began his career at Bemis by observing the morning ritual of opening and sorting the daily mail.

Dick watched as his dad and Carl Jensen reviewed invoices, checks and orders at corporate offices, then located at Plant A on Broadway Avenue in Sheboygan Falls. Each piece of mail was read aloud, sorted into its respective pile and tallied mentally to reveal if bills and payroll could be met that week. Mail was then dispatched with instructions to Executive Secretary Edna (Mickey) Kohlhagen, better known as "Mickey on the spot."

For Dick mail duty was a step up from jobs held during college when he unloaded boxcars of bagged wood flour at Plant B and painted croquet balls at Plant A.

"I really wanted to study business at the side of my father," recalled Dick years later. "I perceived him to be a good businessman. He was my hero."

In late 1964 Dick was promoted to advertising manager by then Sales Manager John Kalk, who introduced him to Bemis sales representatives at the company's first national sales meeting in January 1965. With a freshly printed business card and a title, Dick accompanied John on the road making sales calls to Bemis customers across the country and abroad, as he promoted the family business.

Following the November 1963 assassination of John F. Kennedy, Lyndon Baines Johnson was hastily inaugurated as President while airborne aboard Air Force One. Third-generation Bemis family members Dick and Peter joined the company during LBJ's tumultuous term in office.

Loss and gain

The year 1964 was one of loss and of gain at Bemis. In January a portion of Plant B was damaged when an explosion occurred in a paint tunnel at the finishing operations. The company was still recovering from the loss of major customer Brunswick, who chose to manufacture its own bowling pins and ceased buying from Bemis.

Company President F.K. Bemis (center standing at left) and Sales Manager John Kalk present the Bemis full line of toilet seats to reps at a sales meeting, circa 1965.

Yet 1964 was a year of advances as Bemis produced its first solid-plastic toilet seat. Together with molded wood seats, the plastic seat broadened the company's product line and put Bemis in the same league with industry leaders Beneke and Olsonite.

Molded of HIPS, or high impact polystyrene, the Bemis plastic seat featured a unique and superior high-gloss finish. This was achieved by steam heating the mold before injecting the liquid plastic, which prevented the plastic surface from dulling.

These and other innovations enabled Bemis to gain ground in the competitive plumbing products industry. By 1964 Bemis toilet seats were being shipped to not only all 50 states, but internationally as people worldwide looked to "Be Seated by Bemis."

The compression molding side of the business was also prospering, as Bemis Sales Manager Jere Sampson and his staff touted the advantages of this process to diverse industries. A host of compression molded products—among them croquet balls and mallets, hamper tops, golf tee markers and furniture components—were being brought to market by Bemis. Some flourished and others failed. Yet, through it all, Bemis kept pushing the boundaries of innovation.

1963	1963	1964	1964	1965	1967	1969	1969
Dick Bemis, eldest son of company President F.K., begins his career at Bemis Manufacturing Company.	President John Fitzgerald Kennedy is assassinated in Dallas on November 22.	The Beatles capture a record TV audience of 73 million viewers on *The Ed Sullivan Show.*	The Model 800 is the first injection molded toilet seat made at Bemis.	The Bemis Crandon operation is established in the Wisconsin north woods.	A high-flying counter-culture movement culminates as flower children celebrate the "Summer of Love."	Astronaut Neil Armstrong is the first man to walk on the moon.	Peter Bemis begins his career at the family company, joining brother Dick and father F.K.

Bemis engineer Jeff Trowbridge captured the company's character and strategy in 1964 with these words from a feature story published in *The Sheboygan Press:* "We are actively searching for new areas for our products and for new products for our areas. We are constantly growing and developing. Our all-plastic toilet seat is an example of new development in the industry."

Bemis manufactured an "all-transistor" ship-to-shore marine radio in the 1960s. Featuring the era's "advanced solid-state circuitry," the radio was an attempt to continue to diversify. But the product was short-lived, as Bemis realized the company didn't belong in the radio business.

Woodworking expands to northern Wisconsin

In the summer of 1965 Bemis relocated a portion of its woodworking operation, including all bowling pin manufacturing, to northern Wisconsin. This move would create more room for expansion at Plant A in Sheboygan Falls. Relocating woodworking closer to plentiful hardwood maple sources would also lower production costs.

Bemis found an abandoned plywood plant in Crandon, Wisconsin, located about 175 miles north of Sheboygan Falls. The Streckert Plywood Company had declared bankruptcy in 1961 after just four years in operation. It was formerly the well-known W.B. Gambill Distillery Co., a Forest County fixture that dated back to the days of prohibition.

Bemis agreed to lease the Streckert property from the city of Crandon. It included a brick factory, warehouse, office and dry kilns. In short order Bemis driver Hank Back transported a truckload of planers, gang saws, lathes, glue clamps and other woodworking equipment from Falls to the new north woods operation.

Moving a portion of Bemis woodworking to northern Wisconsin proved a formidable challenge. This was the first manufacturing site removed from company headquarters in Sheboygan Falls.

Additionally, the Streckert plant was in ill repair, and the staff knew little of the fine woodworking skills that Falls workers had mastered for over half a century. Teaching these skills and bringing the plant up to specifications became the challenge of the on-site staff. By early 1967 workers had completed their setup tasks at Crandon.

Bowling pin production continued to grow at the new north woods operation, fueled in part by demand from Japan.

Molcor guitars break the mold at Gibson

Jimi Hendrix, the Who's Pete Townsend and other sixties rock icons may have stomped and set fire to their guitars to energize their onstage performances. But if they had played a Gibson guitar with a body made of Bemis Molcor, Jimi or Pete would have never been able to shatter their stringed instruments.

Super-strong electric guitar bodies were compression molded by Bemis for Gibson in the mid-1960s. It was the era of the Beatles leading the British invasion, a time when every shaggy-haired American teenage boy yearned to crank out riffs like John Lennon and Keith Richards. Guitar sales were soaring, and Bemis—ever seeking new opportunities to apply its compression molding expertise to new applications—looked to jump on board for the ride.

So the company's Contract Sales Manager Jere Sampson approached Gibson, based in Kalamazoo, Michigan, to promote a novel concept—a highly durable solid-body electric guitar made of wood flour and resin. A Molcor body would ensure the guitar would resist breakage, splitting or warpage. It was also less labor-intensive, and thus less expensive, to manufacture than a solid-wood body. But Molcor added considerable weight to the instrument. Unfortunately, that weight contributed to the product's demise. Yet, in the short time that the company made music with Gibson, Bemis enjoyed its contribution to guitar design—and to the growth of rock and roll.

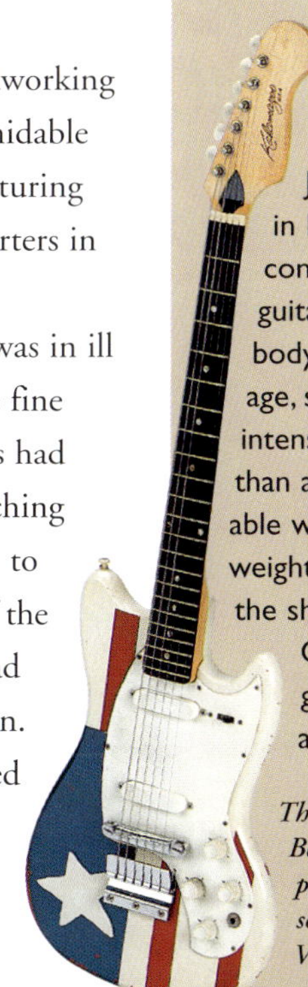

This electric guitar was manufactured by Gibson with a Bemis-made Molcor body. It was personalized with paint by its owner to reflect the political activism and social conscience of the late 1960s and the turbulent Vietnam War era.

The Crandon woodworking operation in Wisconsin's north woods was the former home of W.B. Gambill Distillery Company. Shown here is well-known whiskey brewer C. Everett Beam (center) of Jim Beam fame. The Gambill Distillery closed in 1950 and later became a plywood factory until Bemis moved a portion of its woodworking operation there in 1965.

In the early 1970s the Japanese had discovered bowling, rekindling demand for the sport and the products that supported it—including bowling pins.

By 1973 the Crandon plant was producing 6,800 bowling pins per day—400 more than workers had ever produced during pin production in its heyday at Sheboygan Falls. Many of these pins were shipped to Japan to feed that country's bowling frenzy until the late 1970s when interest waned and Bemis exited the business.

Sporting goods production includes baseball bats

Production of croquet balls, golf tee markers and bowling pins signaled the company's presence in the sporting goods industry in the 1960s. Bemis sought to expand that presence by adding baseball bats.

Looking for other sporting goods products that could use the same gluing skills learned from bowling pin production, company engineer Jeff Trowbridge decided to try to make a better baseball bat.

At this time most bats were made from white ash, a lightweight, yet strong wood. While batters appreciated the reduced weight, they didn't like the fact that the ash bats would crack and break on occasion.

So Jeff replaced ash with hickory, a stronger wood that was less prone to breakage. Bemis even purchased a plant in Martinsville, Indiana, near plentiful sources of hickory. There plant personnel split wood before final manufacturing of Hickory Sticks.

Bemis hoped to create a revolution in baseball with bats that would last twice as long as conventional ash ones. While initial reviews were favorable, it wasn't long before the benefits of traditional ash outweighed the new alternative, and Bemis hickory bats were scratched from the company's product line.

Hickory Sticks with holes

Inmates at Sing Sing, the historic maximum security prison in upstate New York, preferred Bemis Hickory Sticks for their pickup games of baseball. So did California college athlete Al Diamond, who claimed in 1965 that the bat transformed him from a mediocre hitter to one of college baseball's best sluggers. Even the eminent *Wall Street Journal* lauded the advantages of the Bemis Hickory Stick and its inventor, company engineer Jeff Trowbridge:

"The next time a baseball fan accuses a strikeout victim of having holes in his bat, he could be right . . . (Bemis) invented and now sells bats with holes drilled in them," wrote the *Journal* in July 1968.

Just as Bemis laminated bowling pins had holes drilled in their inner core to adjust their weight, company woodworkers drilled holes through the thick barrel of hickory bats to lighten them and approximate the weight of ash. Holes created by this patented process were covered with glued laminated "cheeks" before the bat was turned and finished.

Not one of the ultrastrong Bemis Hickory Sticks ever broke in use. But pity the poor batter who connected with the ball because for all of its advantages, the Hickory Stick vibrated like a tuning fork on contact with a baseball. Unfortunately, the Hickory Stick's reputation as a stinger relegated this bat to the dugout. The product was dropped from the Bemis line in 1968, a few years after it was first introduced.

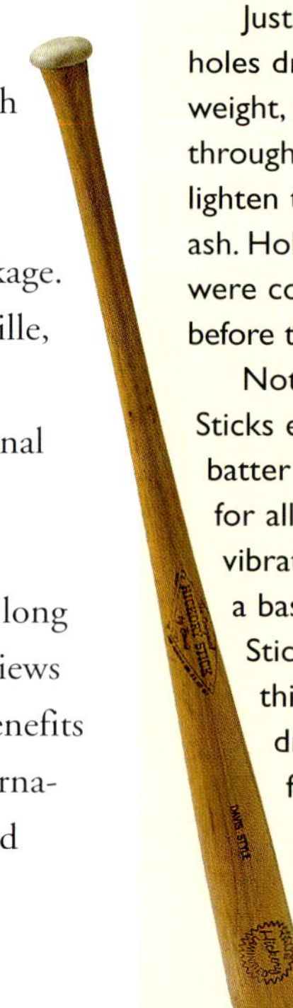

The Bemis Hickory Stick was a stronger bat that resisted breakage, but it resonated on contact with the ball. The bat's reputation a stinger contributed to it being ousted fro the Bemis product lineup.

Area workers install footings for a $100,000 addition to Bemis in December 1968. The footings supported precast concrete walls to create more space at the company's Plant B site on Mill Street in Sheboygan Falls.

New buildings signal growth

As business expanded in the mid-1960s, Bemis was fast outgrowing its office space. Located at the old Broadway Avenue plant (Plant A), the office area was cramped and confining. Moving to larger quarters became a priority.

In 1968 a new office area was added on to the expanding factory site known as Plant B. This was the parcel of property on the western edge of Sheboygan Falls—the former Heronymus farm acreage with the pea-canning factory that Bemis had purchased in the late 1940s. Bounded on two sides by the Sheboygan River, the new office area offered more room than Bemis staff thought they could ever use.

Yet, just a year later, Bemis again needed to expand to keep pace with the company's growth. So Bemis launched a construction project to add 25,000 square feet of floor space adjacent to the office building. The new area, slated for manufacturing and warehousing, was completed in the spring of 1969.

Gas station giveaways bring small-parts molding to Bemis

They were part and parcel of every all-American family's kitchen in the plastic and prosperous 1960s. Durable and colorful bowls, mugs and dishes in avocado green, harvest gold and other now-dated hues—products that were free with every fill-up at the neighborhood service station. Their manufacturer was Bemis.

In the mid-1960s Bemis looked to expand its Contract Injection Molding business by manufacturing a variety of products for a host of original equipment manufacturers (OEMs). Among them was David Douglas & Co., Inc. of Manitowoc, Wisconsin.

David Douglas distributed plastic tableware to service stations nationwide as incentive premiums. With gasoline prices low and station owners waging price wars for motorists' business, giveaways were an inducement to build customer loyalty and lure new patrons.

Manufacturing plastic tableware helped Bemis master its injection molding expertise, as the company developed new molds and the technology to create a variety of small plastic shapes.

The alliance with David Douglas & Co. was so successful that Bemis purchased additional injection molding machines to keep pace with product demand.

Plastic salad and snack sets, coffee decanters shaped like genie bottles and insulated coffee mugs sporting gas station logos were among a host of products that Bemis molded for Wisconsin-based David Douglas & Co. in the 1960s.

A 1960s era product sales flyer describes Bemis toilet seat features and benefits. The Bemis Model 1950, a seat similar to this one but with a lid, was shipped by the thousands to U.S. military facilities in Southeast Asia during the Vietnam War.

Bemis seats in Vietnam

Just as Bemis had supplied soldiers in World War II and the Korean War with a much-needed essential—toilet seats—the company also shipped its products to the steamy jungles of Vietnam.

During the Vietnam conflict, global construction firm Morrison Knudsen, or MK, was contracted by the United States government to build landing strips, barracks and other military facilities in Southeast Asia. MK purchased toilet seats for those quarters from Bemis.

Over a two-year span Bemis packed thousands of commercial elongated plastic seats in wooden crates destined for the other side of the globe. And American combat forces waging war against Communism found they could "Be Seated by Bemis" in places as remote as Southeast Asia.

What's my line?

Bemis Vice President of Sales Carl Jensen and wife Lucille ham it up during a spoof of the popular 1960s TV game show, *What's My Line?* Carl and Lucille sport toilet seats slung around their necks and a sign to entice the audience, while on vacation aboard a 1960s cruise ship. Carl retired in 1972 after 40 years with Bemis. His tenure overlapped three generations of ownership—beginning in 1932 with company founder Albert Bemis until Al's death in 1947, Al's son F.K., and current third-generation leadership under Dick and Peter.

Peter Bemis joins the company

The Vietnam War was at full throttle in 1969, as antiwar rallies raged on campuses nationwide, and a generation came together for rock music, drugs and peace at a farm outside of Woodstock, New York.

Peter Bemis, 1969

It was during that riotous year that Peter Bemis, youngest son of company President F.K. Bemis, graduated from Wisconsin's Carroll College. Peter had earned a bachelor's degree in business administration and economics in June—and a low number in the military draft lottery in December.

Rather than place his rank, assignment and fate in the hands of the draft, Peter chose to enlist in the Army's OCS (Officers Candidate School) to gain more control of his destiny. But high blood pressure barred him from gaining admission to OCS.

So Peter directed his attention to the business in Sheboygan Falls instead of heading off to war in Southeast Asia.

"I had a really strong desire to work for the company," recalled Peter years later. "I think it went back to growing up as a kid and hearing my father talk about the business. My dad was really brilliant, just a genius. He was also one of the few businessmen I knew who had balance. He had all the core values down—work hard, but there's no purpose in succeeding if you can't do it in an honest way. He had a strong moral ethic."

From desk job to plant job

Peter's first assignment at Bemis was as an efficiency report clerk earning $2.25 an hour. But he was restless in this desk job and often wandered out into the plant after hours to watch press operators in action. It was in the plant that he discovered his passion for manufacturing.

"I'd work on setup. At night I'd go out and make adjustments to machines," said Peter. "The foremen could always use a hand. At times I think I probably drove them a little crazy. But I didn't want to be a sales guy. I always liked to investigate how things worked."

Peter's father soon recognized his son's affinity for manufacturing. It wasn't long before Peter vacated his desk job and assumed new duties as a manufacturing manager in the company's Injection Molding area. It was in that role that he would guide Bemis to becoming one of the nation's premier high-technology injection molders.

The 1970s and early 1980s brought a host of "firsts" to Bemis, including acquisitions and new products for new industries. Supporting this growth was the company's core product line—toilet seats—including compression molded models shown here. Suspended from hooks on a conveyor, seat components moved through automated paint tunnels where the Bemis flow coat enamel finish was applied. This process, developed in the 1950s by company engineers, is still in use today.

CHAPTER 6

An era of "firsts"

It was the time of Watergate and the dishonorable fall of a U.S. President. An embargo of Arab oil-exporting nations was causing chaos at gas stations across America. In the volatile Middle East militants seized the U.S. Embassy in Iran, taking 90 Americans hostage for 444 days.

Following the move of Wisconsin Plastics to new facilities, a portion of Bemis Plant A was razed. The demolition of the aging building helped make room for the revitalization of Sheboygan Falls.

Back at home disco dancing was the latest rage, inspired by a leisure suit-clad John Travolta spinning under strobe lights in the Hollywood film *Saturday Night Fever*. And a revolution in office automation dawned as IBM and an upstart technology company called Apple introduced the first "personal computer."

In Sheboygan Falls Bemis was continuing to build on the success of its earlier years. The company had carved out a solid place in the U.S. plumbing products area and a niche business in injection and compression molded products for diverse industries. Through these successes Bemis had built enough of a balance sheet to expand by acquiring other firms—one of many "firsts" this era would bring to Bemis.

Wisconsin Plastics is first BMC acquisition

It was business ingenuity and good timing that brought Bemis its first acquisition since the company's predecessor, Bemis-Riddell Fibre Co. and its co-owner Albert Bemis purchased White Wagon Works back in 1924.

The year was 1971 when Bemis management acquired Wisconsin Plastics, a Sheboygan-based contract extrusion firm. The purchase was significant for several reasons. One, it was the first acquisition for Bemis Manufacturing Company. Wisconsin Plastics also brought a whole new process to Bemis—thermoplastic profile extrusion. The acquisition signaled the company's continuing commitment to diversify not only its products but its expertise in plastics processing by adding extrusion to its compression and injection molding capabilities.

Bemis moved Wisconsin Plastics from Sheboygan to the basement of Plant A in Sheboygan Falls. One of the new division's first products was a part for the U.S. automotive industry—a J-shaped retainer that held car seat cloth in place. Other products included profiles for office furniture and sight chambers for the medical industry.

While the Wisconsin Plastics name and logo were eventually discontinued, the acquisition, now known as BCE (Bemis Contract Extrusion), remains a vital division at Bemis Manufacturing Company.

Bemis engineer Dan Cykana explains manufacturing at the company's Wisconsin Plastics Division to plant visitors at a 1980 open house. Shown at far right is Dan's colleague Henry Dentz.

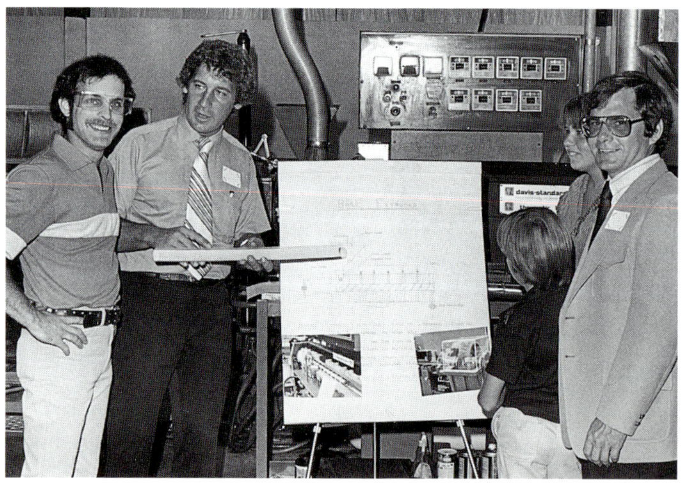

Farewell to Carl Jensen

From the day in 1932 that he convinced Al and Oleida Bemis to hire him to his retirement 40 years later, A.C. (Carl) Jensen was instrumental in the growth and success of Bemis Manufacturing Company.

Formerly with the Crocker Chair Company of Sheboygan, Carl was out of a job when the Great Depression brought bankruptcy to Crocker. But he retained Crocker's customer list—and sold that list plus his expertise and business connections to Al Bemis in exchange for a job and 25 percent company ownership. It was the start of Bemis Manufacturing Company as we know it today.

Carl led the sales function at Bemis until he retired in 1972. He also held corporate offices as vice president, secretary, treasurer and part-owner. But it was his astute and sound business judgement that brought the most value during his 40-year career at Bemis.

1972	1973	1973	1975	1976	1977	1977	1982
Cigarette advertisements are banned from the nation's airwaves.	Carl Rickmeier II replaces Carl Jensen as new vice president of sales and marketing at Bemis.	A beleaguered President Nixon declares, "I'm not a crook" in the midst of the Watergate scandal.	Bemis acquires a new 2,000-ton press to support the growing West Bend business.	Bemis introduces red, white and blue bedpans and colonial-designed toilet seats for America's bicentennial.	A new flow coat line at Plant B doubles Bemis finishing capacity.	Two huge extraterrestrial hits, *Star Wars* and *Close Encounters of the Third Kind*, debut to moviegoers nationwide.	Bemis introduces a line of plastic rain gutters and downspouts under the Rain Master brand.

Growth at Contract Injection Molding

The purchase of Wisconsin Plastics signaled the start of a successful Contract Extrusion operation at Bemis.

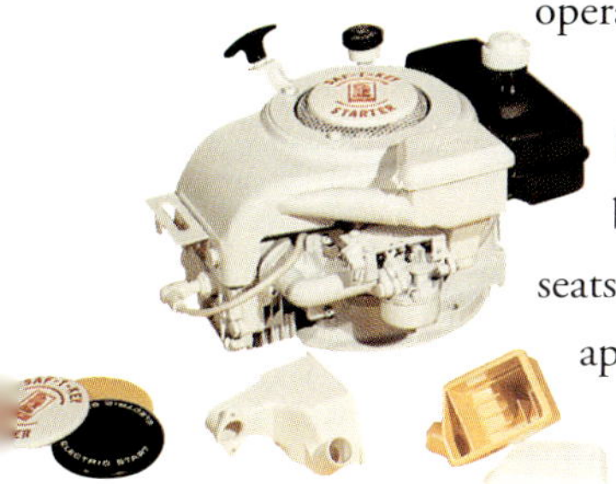

Small engine components were manufactured in the 1970s by Bemis Contract Injection Molding. These OEM parts were made for Lauson Engine of New Holstein, Wisconsin, a supplier to Tecumseh Products Company.

But at Injection Molding business was stalled, in part, because demand for plastic toilet seats, sold primarily for commercial applications, was stagnating. The division had few accounts—Tecumseh engine parts for Lauson Engine of New Holstein, Wisconsin; hamper components for Harvey Manufacturing Company of Cary, Illinois; and gas station incentive premiums for David Douglas & Co., Inc. of Manitowoc, Wisconsin.

To acquire new customers and increase profitability at Injection Molding, Peter Bemis joined with Sales Manager Steve Larson and Engineer Jerry Swart to strategize growth for the division. Their efforts and those of their coworkers brought Bemis new volume, new opportunities and new equipment to pursue those opportunities.

During the 1970s Contract Injection Molding manufactured a host of diverse products for companies scattered throughout the Midwest. These included cafeteria table seats and backs, carpet sweeper brush blocks, point-of-purchase signage for Hallmark and Coca-Cola, baby crib components, envelope moisteners and tissue dispensers. Many of these were formerly made of wood, ceramic or metal. But Bemis recognized that solid plastic was a more cost-effective alternative for customers—and offered more opportunity for the company.

Bemis Contract Injection Molding manufactured diverse products for many industries during the 1970s. These included clothes hamper tops, rims and other components for Harvey Manufacturing Company of Cary, Illinois.

While Injection Molding expanded its customer base during the 1970s and early 1980s, the division also increased its technology and expertise. This set the stage for large press molding and processing of engineered materials, a critical evolution for Bemis that was yet to come.

In the 1970s Bemis molded a colorful one-piece body for the Garton Toy Company Chubby Car. It was a return to the company's roots in children's wheeled toys, as Bemis predecessor White Wagon Works made wooden coaster wagons at the turn of the twentieth century.

Oil embargo signals end of gas station giveaways

The successful injection molding business that Bemis had built during the late 1960s making tableware incentive premiums for retail gasoline companies came to a screeching halt with the 1973 oil embargo.

It was the onset of the worst energy crisis since World War II—a time of long lines at the gas pumps, limited gasoline sales and a 55-mile-an-hour national speed limit.

The fuel shortage was caused by an oil embargo declared by the Organization of Petroleum Exporting Countries. OPEC launched the gas war to gain control over petroleum consumption in the U.S., Japan and other major industrialized countries.

No longer did gas station owners have to lure customers to their pumps with offers of free plastic tableware for every fill-up. Virtually overnight the Bemis plastic premium business with David Douglas & Co. evaporated.

Plastic salad bowl sets and avocado green thermoinsulated tumblers like these were no longer viable giveaways when a gasoline shortage struck the U.S. in 1973. Their demise brought an end to the Bemis business with David Douglas & Co., a Wisconsin-based firm that distributed these products as incentive premiums to gas stations nationwide in the 1960s and early 1970s.

Launched in 1972, in part, to replace the lost gas station giveaways business, Bemis Health Care, or BHC, produced plastic products for hospitals and health care facilities nationwide.

Birth of Bemis Health Care

With the loss of the gas station premium business, Bemis was presented with a new opportunity—molding disposable products for the health care industry. The birth of Bemis Health Care, or BHC, gave the company an enormous boost when it needed it most.

Carl H. Rickmeier II

In 1971 Sheboygan businessman Carl Rickmeier approached Bemis with an idea on how the company could fill idle press time. Carl recognized that plastic was superior to stainless steel for bedpans, washbasins and other hospital supplies. It was a time when sterilized stainless steel was the material of choice for health care facilities. But disposable plastic offered advantages, according to Carl—including lighter weight, better infection control and lower cost.

Health Care business takes off

In response to Carl's pitch Bemis created a new joint venture and called it Bemis Health Care. The company contracted with a Detroit industrial design firm to develop a hospital "admission kit."

Comprised of a washbasin, water pitcher, drinking cup, serving tray, emesis basin and other personal hygiene items, the admission kit was color-coordinated and practical with pieces nesting together for easy storage.

Carl marketed admission kits to health industry giants, including American Hospital Supply and Sherwood Medical. Not only were they interested, so were other companies. Within one brief year Bemis added a 45,000-square-foot addition at Plant B and purchased additional injection molding machines to keep pace with demand for its health care products. These would eventually include suction canisters used in surgical procedures and sharps containers for safe disposal of used needles and syringes.

Western property purchased

With health care items, contract products and toilet seat sales boosting Bemis Injection Molding operations, company management recognized the need to expand beyond the property at Plant B. So in April 1973 Bemis purchased land on Sheboygan County Trunk Road PP just west of Sheboygan Falls.

Formerly a farm owned by the LeMahieu family, the land parcel was 80 acres in size and gave Bemis room to grow, unencumbered by the river and railroad line that limited further Plant B expansion.

Within one year a new 100,000-square-foot warehouse—Plant C—was opened at the new western campus. As further growth became necessary, Bemis purchased adjacent

Expansion to Bemis corporate headquarters was one of many building projects at Bemis in the 1970s and 1980s as the company continued its growth and success.

Top-Tite hinge propels Bemis to the top

The task of switching a toilet seat shouldn't be a difficult one, especially if Bemis were to sell replacement seats in large quantity. But because of corroded bolts and hinges, changing a seat was often quite an ordeal.

The importance of a corrosion-free toilet seat hinge became apparent to company Vice President Dick Bemis in 1970—the year he and his family moved to a different home in Sheboygan. Dick attempted to replace all the toilet seats but found their steel studs corroded, making removal nearly impossible.

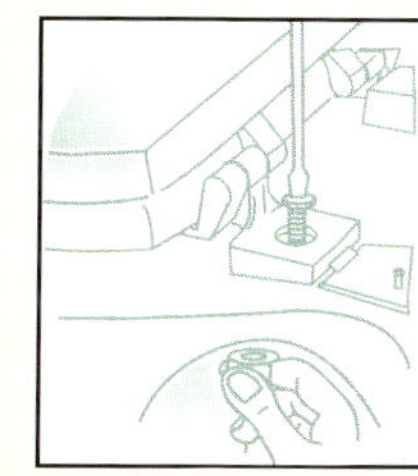

He was forced to use a hacksaw to dislodge the bolts, and it was in his frustration that Dick realized Bemis had to offer a better hinge to expand sales. The result was the Top-Tite hinge.

A watershed in the company's history, the Bemis Top-Tite hinge boasted unique nylon rust-free hardware. It made competitors cringe and customers cheer. Best of all it created attention as other industry players recognized Bemis as a significant competitor.

While the Top-Tite was eventually imitated by competitors, Bemis had gained sufficient market share by that time to be recognized as a leader and innovator in the plumbing products industry.

acreage from the nearby Bimmel farm and additional property from scattered nearby parcels. This location would eventually grow to over 300 acres as Bemis continued to expand.

Crandon launches cutting boards

During the 1970s the Bemis woodworking operation at Crandon was in a state of flux. Bowling pin production, peaking at 6,800 pins per day in 1973, was falling off as interest in the sport waned.

Because pins were a contract product, they were subject to price and market fluctuations over which Bemis had no control. Crandon could no longer rely on pins to support their north woods operation. So in 1974 Bemis brought another product to Crandon—wooden cutting boards. The new product line signaled the birth of Bemis Housewares.

Sold under the Colonial Kitchen brand, cutting boards brought diversity and stability to Bemis woodworking operations. The company eventually added chopping blocks, bread and recipe boxes, spice racks and canister sets.

Colonial Kitchen.

During the 1970s Bemis expanded its woodworking operations at Crandon, Wisconsin, to include a host of new housewares items including bread boxes like this one, marketed under the Colonial Kitchen brand.

Shut Box game wins fans far and wide

Perhaps no other Bemis product from years past evokes such endearing memories as the Shut Box game. For three decades thousands of families have enjoyed hours together playing the unique dice board game.

Shut Box was first discovered by F.K. Bemis on a vacation to the Caribbean. Considered a gambler's pub game, Shut Box was comprised of a wooden felt-covered board, dice and numerals behind wooden slats. The novel game caught F.K.'s interest, and he returned to Sheboygan Falls with one for personal use.

It wasn't long before F.K.'s friends and family discovered Shut Box and requested copies of their own. So Bemis Woodworking Foreman Pat Babb duplicated the game in his home wood shop, and assembly and finishing were completed at Plant A.

Bemis made about 30,000 Shut Box games from the 1950s through the 1980s, when the company discontinued its production. Yet, even today, the Bemis Shut Box remains a treasured memory for families far and wide.

Large press molding comes to Bemis

When customer West Bend chose to make a large one-piece plastic humidifier body, the company looked to Bemis.

In 1974 Bemis Contract Injection Molding won a new contract with West Bend, one that would set the company in a whole new direction—a progression to large press injection molding.

To manufacture the new humidifier body, Bemis had to purchase a press bigger than the company had ever owned—a new 2,000-ton injection molding machine. Dubbed the "green monster," the 2,000-ton Cincinnati Milacron press was shipped in portions and required a special multiwheeled car for safe transport over area railroad bridges.

Led by Peter Bemis, plant engineering staff Jerry Swart, Jack Nelson and Jeff Trowbridge developed part design and tool construction techniques to manufacture the humidifier body for West Bend. Their sophisticated basket tool had many moving parts with all four sides opening around the core. Its development propelled Bemis forward in tool engineering expertise.

The arrival of the 2,000-ton press in 1975 marked a turning point in the company's six-year relationship as a supplier to West Bend and led to Bemis eventually purchasing the humidifier division eight years later.

Instrumental in the development of large part molding expertise at Bemis were Peter Bemis (right) and company engineer Jack Nelson (left). Together with other manufacturing, sales and engineering staff, Peter and Jack developed a one-piece plastic humidifier body for West Bend. They are shown here with Bemis retiree Don Hailer.

Furthermore, it significantly advanced the company's engineering capabilities. More important, the 2,000-ton press and the West Bend business that it supported signaled the genesis of the sophisticated contract molding capability that Bemis is widely known for today.

Large press molding—the Bemis advantage

Large press molding. It's a capability many other injection molders shun. Molding parts on large presses takes significant investment in machinery, floor space, employee training and safety procedures. Large press molding is not for every plastics processor. But over the past quarter century Bemis has mastered molding on oversized injection presses—and using challenging engineering grade materials instead of easy-to-mold commodity plastics. That expertise has enabled the company to work with industry leaders and Fortune 500 companies nationwide.

Workers unload a new Cincinnati Milacron 2,000-ton press at Bemis in 1975. This injection molding machine was the largest ever acquired by Bemis to date and signaled the onset of the company's expertise in large press molding. Today that 2,000-ton machine is dwarfed by the company's latest press, a 6,600-ton co-injection machine.

Stasco acquisition spurs seat sales

It was a former sewing factory for military uniforms during the Civil War. Aged, in ill repair and located in a declining inner-city neighborhood in Camden, New Jersey. Yet Stasco, or Standard Tank and Seat Company, represented a major acquisition for Bemis.

In 1975 company Vice President Dick Bemis, Treasurer Jim Kroscher, Legal Counsel John Howell, and Sales Manager John Kalk traveled to Camden to negotiate the Stasco deal. They were greeted by a crumbling plant surrounded by a chain-link fence and a vicious attack dog guarding the property. Despite these obstacles Bemis finalized the purchase which included the Stasco brand, toilet seat inventory and customer accounts.

The Stasco acquisition—only the second in the 74-year history of Bemis—was a significant strategic move. Never before had the company acquired a competitor in the plumbing products industry. Stasco also brought Bemis many new East Coast customers and helped the company expand seat sales on a region-by-region basis.

A Mayfair retail product catalog from the 1970s shows the wide array of Molcor toilet seat styles with bright enamel finish colors and lively lids. Styles included a fanciful frog, winking owl and American eagle to honor the nation's 1976 bicentennial.

A Bemis machine operator removes a bathroom vanity from an injection molding press. Manufactured for a brief time in the mid-1970s, these sinks were targeted for mobile homes and other select markets.

The faces of Bemis

The 1970s brought an employee boom to Bemis as staff swelled to keep pace with the company's growth. Most of these new hires were production workers.

Since the company's inception, plant employees have steered Bemis to success with their skills, teamwork and pride in their jobs. Shown at right are five workers in the 1970s. Some are retired and others are still employed at Bemis. They are just a few of the thousands of hourly employees who helped to make Bemis all that it is today.

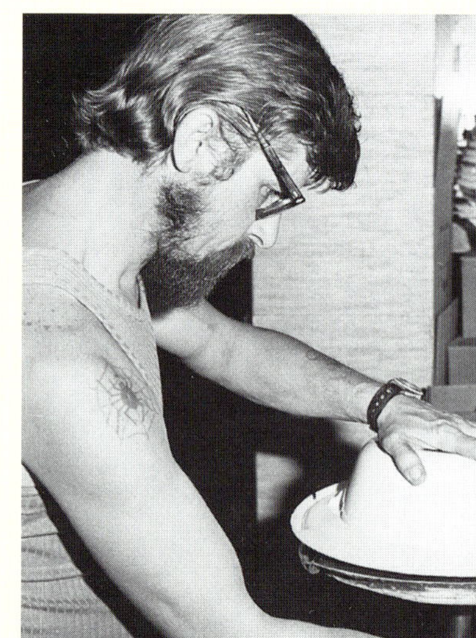

Tom Malwitz

Rain Master brings outdoor home improvement to Bemis

Extrusion and Injection Molding came together in 1982 as Bemis launched a new product line—rust-free plastic rain gutters and downspouts. These were marketed under the new Rain Master brand.

RAIN MASTER
ALL-WEATHER VINYL GUTTER SYSTEM

Driving the effort was company Manager of Product Development Ken Pepper. Ken's process engineering expertise in polyvinyl chloride, or vinyl plastic, was indispensable as Bemis had never molded this material before.

In late 1981 Bemis purchased two new Cincinnati Milacron CM-80 twin screw extruders to produce downspouts and gutters. They were delivered to the company's Wisconsin Plastics Division, now located at the western campus in the new Plant D facility.

Sold through major hardware chains nationwide, the Rain Master line leveraged the company's existing relationship with these retailers, many of which already purchased Bemis products under the Mayfair and Colonial Kitchen brands.

The addition of Rain Master and the company's Home Improvement Division broadened the Bemis product line. It also continued the company's commitment to enter new and viable industries, and re-engineer itself to keep pace with ever-changing technology and market opportunities.

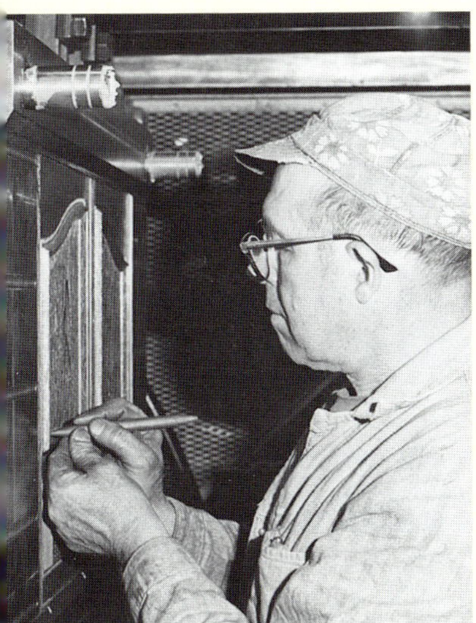

Ben Ryder

Gene Soerens

Marion Weiler

Ted Lueck

Bemis Sheboygan Falls machine operator Bettymae Schuh poses for a photo in the M&I Bank 1994 annual report. The plastic products surrounding Bettymae illustrate Bemis world-class skill in large part molding of both commodity and highly engineered thermoplastics for diverse applications—from lawn mowers to toilet seats and resin furniture to tractor fenders.

CHAPTER 7

Acquisitions fuel growth

They've been called the Reagan years. The 1980s were a time of Gorbachev and *glasnost*—or Russian for "openness." It was an era of optimism when the Berlin Wall was felled, and a high technology explosion revealed a brave new world of information.

But the 1980s were also a time of excess and cynicism. Wall Street was plagued by insider trading. Pete Rose was banned from baseball for life for allegedly placing bets on ball games. And the space shuttle *Challenger* exploded in a shocking television scene witnessed by millions.

In the business world, mergers, acquisitions and corporate restructurings were transpiring as companies looked to expand market share, create efficiency and position themselves in a rapidly evolving global economy. Bemis was no exception.

From 1983 to 1994 Bemis added eight firms to its expanding portfolio—more than it had acquired since its inception. Half were Bemis competitors. The others were purchased to diversify company product lines.

"The Best Seat in the House" is what Church has been known for since its founding in 1898. Bemis purchased Church Seats in 1983 as the company looked to expand market share and its core toilet seat product line.

It was an exciting time at Bemis—and third-generation company heirs Dick and Peter were leading the charge.

Bemis acquires Church Seats

The year 1983 brought two major acquisitions to Bemis. The first one was Church Seats Company of Monson, Massachusetts, purchased to expand Bemis East Coast market share in toilet seats.

A prominent and well-respected brand in the plumbing products industry and a former division of American Standard, Church was known as "The Best Seat in the House."

Bemis management had been in negotiations with Church owners for several years before the purchase was finalized in 1983. With the acquisition Bemis obtained rights to the "Best Seat" tag line, the Posturemold name, and multiple patents on toilet seat and hinge design.

The Church purchase was important for Bemis. It brought more production volume to Sheboygan Falls and broadened the company's core seat business. Church also expanded the Bemis customer base to include premier East Coast wholesalers and added a new crew of knowledgeable sales representatives.

Church®
The Best Seat in the House®

Final inspection and trimming of a plastic injection molded toilet seat ring is carried out by employee Randy Mills. This photo appeared in the first Bemis corporate brochure, produced in 1989.

The acquisition also brought the much-lauded Church name to the Bemis stable of brands—one that Bemis would set apart as the premium brand for the specifying engineer and mechanical wholesaler. Today Bemis continues to support and leverage the Church name to the plumbing products industry.

The comical "Mr. Dry Air" caricature was developed for Bemis in the 1980s. During these years Bob Claus (shown at left) was sales manager for the company's Humidifier Division.

West Bend purchase signals start of Home Comfort Division

Church Seats was not the only business acquired in 1983. In October of that year Bemis signed an agreement to purchase the humidifier division of West Bend. The move was a strategic one to preserve existing business, as West Bend was seeking to sell its humidifier division. Bemis Contract had supplied plastic components for West Bend humidifiers and other appliances since 1973.

With the acquisition Bemis entered a new industry—small appliances—and launched the Waterwick Humidifier Division, the predecessor to Bemis Home Comfort. Over the next three years West Bend gradually transitioned the business as Bemis gained engineering, manufacturing and marketing expertise in the appliance industry.

Bemis hired industry veteran Robert Claus to manage sales efforts, and consumer advertising was launched to promote humidifier products to a nationwide audience. Part of this promotion included "Mr. Dry Air," a whimsical whiskered caricature that proved a hit with customers.

1983	1984	1984	1985	1985	1987	1991	1992
President Ronald Reagan declares the Soviet Union an "evil empire."	A new Bemis R&D facility opens near Plant B, with Ken Pepper as manager.	Wisconsin Plastics Corporation is dissolved on December 31. It is renamed Bemis Contract Extrusion.	An all-star intercontinental rock concert dubbed Live-Aid is staged to raise money for African famine relief.	Bemis purchases Minnesota-based Tommer Cutlery to complement its line of cutting boards.	The stock market plunges 508 points on October 19, wiping out more than $500 billion in stock equity.	Bemis adds a 150,000-square-foot warehouse at its west campus in Sheboygan Falls.	At age 46 Arkansas Democratic Governor Bill Clinton is the first baby boomer to reach the White House.

Gamco acquisition brings new technology

With the Stasco and Church acquisitions Bemis found it could effectively expand market share in toilet seats. In 1985 Bemis applied that same strategy to build its health care business with the purchase of Gamco.

A sister company of Delta Faucets, Gamco was a Kentucky-based manufacturer of plastic patient care products for the health care industry. Gamco also produced a dietary line of plastic cutlery and urethane foam-insulated tableware for hospital food service applications. Beyond health care Gamco manufactured plastic drinkware for TWA, Pan AM, United and other major players in the airline industry.

In the mid-1980s scores of plastic utensils were molded using equipment from Gamco. Shown here are Bemis operators Gail Wensink (foreground) and Barb Sipiorski hard at work.

With the Gamco purchase Bemis made key gains in health care and industries beyond. The proprietary stack mold Gamco had developed was now owned by Bemis. The company moved the mold to Sheboygan Falls and used it to increase production of patient care products—thereby reducing costs. This helped solidify the Bemis Health Care relationship with American Hospital Supply, a key former customer of Gamco.

Insulated tableware for hospital food service applications became part of a new dietary product line at Bemis with the 1985 acquisition of Gamco. Bowls, plates and cups like those shown above were comprised of a twin-wall plastic skin surrounding a temperature-insulating urethane foam core.

In addition to this mold Bemis acquired technology from Gamco that included high-speed molding of thin-walled disposable plastic drinkware and cutlery.

But to keep pace with rapid changes in the industry, Bemis would have had to invest in new technology, as competitors began migrating to vacuum forming. This process created products that were less expensive and more durable than the brittle, thin-walled styrene plasticware that Bemis manufactured.

Gradually, over a two-year period, Bemis lost its high-volume airline drinkware business to this new technology. But by then Bemis had come to realize its future was not in this low-margin, highly competitive business.

Yet the Gamco acquisition taught Bemis a valuable lesson: It's not enough to keep pace with the competition. You must remain one step ahead, or you'll be eclipsed by them and by ever-changing technology.

During the 1980s Bemis Contract Extrusion, formerly known as Wisconsin Plastics, created a variety of plastic profiles in a host of shapes, materials and colors for an array of applications.

Bemis helps plastic bend to new applications

"I just want to say one word to you, just one word, Ben . . . plastics. There is a great future in plastics."

It was abject advice uttered to recent college graduate Dustin Hoffman in the 1967 hit Hollywood film *The Graduate*. To the idealistic baby boomer generation the word "plastics" quickly became synonymous with middle-class sellout. This reinforced plastics' image as synthetic and disposable.

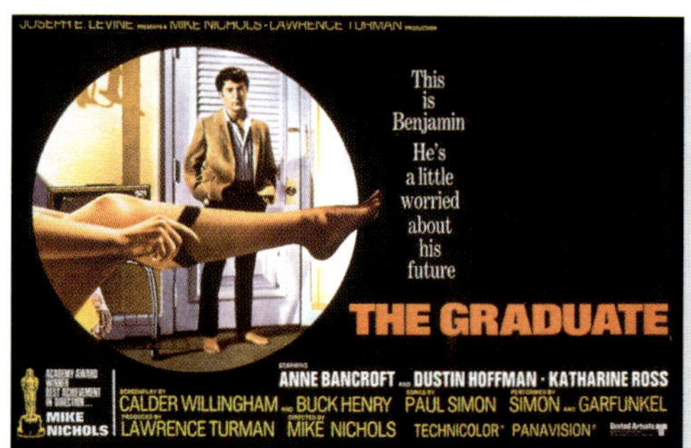

Since its increase in popularity during the the post-World War II years plastic was deemed appropriate for common products like food containers, toys, washbasins and toilet seats. Using this mundane material in more challenging applications was out of the question, in part because early plastics didn't have the stability to withstand intense heat, ultraviolet exposure or impact.

But by the 1980s scientists were developing a new breed of cutting-edge polymers for more rigorous applications, and companies like Bemis were handed the challenge of processing them. Many of these were designed to replace steel or aluminum devices with plastic components that were less expensive, lighter weight, resistant to denting or rusting and didn't require painting. Thermoplastic parts also offered design flexibility never before available with more traditional materials.

The opportunities were enormous for companies like Bemis that recognized the potential of plastics in thousands of commercial products—from cars to computers and telephones to trash bags. It was the start of a scientific revolution with matter being manipulated molecule by molecule. And Bemis was part of that revolution.

Changing of the guard

The 1980s brought many personnel changes to Bemis with the retirements of seats Sales Manager John Kalk, Engineer Jack Nelson, CFO Jim Kroscher and Contract Sales Manager Jere Sampson. But the most significant of these was in 1986, when F.K. "Pete" Bemis stepped down as chairman of the board.

In his 50-plus years of leadership at Bemis Pete brought strength, wisdom and business insight to the company that bore his name. Taking the reins of the business in 1947, upon his father Al Bemis' death, Pete built the company to become a diversified world-class plastics processor and manufacturer.

Sons Dick and Peter, and daughters Patty, Cathy and LuLu all recall the warm and approachable side of their father. Employees concur that Pete was someone who truly appreciated the contributions of others in making Bemis a success. For most of his 50 years at Bemis Pete would tour the plant daily to observe operators at work and chat with them about their lives and their jobs. He knew everyone on a first-name basis. And he listened when employees would voice an opinion about a task or share an idea for how to improve an operation.

It was this nonautocratic, inclusive collaboration between upper management and frontline workers that spelled success for Bemis. That success is evidenced in historically low employee turnover at Bemis and a high number of second- and third-generation families who earn their livelihood at the company.

Pete was a calculated risk taker, and after giving a decision sufficient consideration, he didn't hesitate to make changes that would improve company performance or profitability.

Among these was Pete's decision in the 1930s to dispose of the company's novelty furniture business and instead focus resources on a more viable and profitable product—toilet seats. That decision created a strong core business that has propelled the company to leadership in the plumbing products industry.

"He knew everyone by name," recall employees who worked at Bemis while owner F.K "Pete" managed the troops. Among these was long-time Bemis driver Hank Back (left) celebrating his retirement in 1983 with Pete.

Pete's vision also enabled Bemis to embrace changing technology. In the early 1950s Pete, together with brother-in-law Jack Nelson and cousin Walt Trowbridge, brought new compression molding technology to Bemis. That novel process used pulverized wood flour combined with plastic resin as a binding agent. Not only were these Molcor toilet seats and other products less costly to manufacture than wood, they were more durable and warp-resistant.

As use of plastics skyrocketed in the 1960s and 1970s, Bemis again adapted, adding profile extrusion and injection molding technology under Pete's direction. By the time he retired in 1986, Pete had built his father's company from a small Wisconsin woodworking firm into an international supplier to a host of industries.

Pete passed away in December 1987, but his legacy lives on at Bemis in the products, processes and people of the company that bears his name.

In 1984 F.K. "Pete" Bemis (center) addresses an audience at a banquet celebrating his 50th anniversary with the company while sons Peter (left) and Dick look on.

Quest for quality leads to new ADF manufacturing

To boost product quality and stave off competition from Japan and low-cost overseas competitors, Bemis adopted a new approach to manufacturing in the late 1980s. It was one that featured complex robotics, computers and sophisticated conveyors to create a defect-free product. It was driven by Executive VP of Manufacturing Peter Bemis.

Bemis Vice President of Sales and Marketing Steve Larson played a key role in securing new business to expand the Contract Division, its ADF initiative and technological capabilities.

The new process was dubbed "ADF" or "Automated Defect-Free." The name was an extension of the "defect-free" (DF) statistical methods that Bemis had implemented as part of the company's early problem-solving efforts.

Over two years in the planning, testing and validation stages, the Bemis ADF process was the result of a team effort. This included all phases of production—product development, engineering, manufacturing, setup and management. ADF optimized many process variables—including air pressure, injection pressure and speed, water temperature and pressure. It also alleviated worker fatigue by using robots to perform repetitive tasks and maintain consistency in the machine cycle.

"If Bemis wants to grow and stay competitive and continue to provide its employees with a decent living and benefits, it must make its products without defect or return," said Peter Bemis in the November 1986 issue of the *Bemis Extra Heavy* employee newsletter. "Measuring statistically will help produce a high-quality product in the least amount of time. It's the way of the future."

Colorful plastic products molded by Bemis Contract operations in the 1980s included an assortment of components for a wide range of OEM applications—including a dispensing tube for Procter & Gamble toothpaste (page center).

The Bemis Automated Defect-Free Team, September 1988. They are (from left) Lori Allmann, Steve Kolste, Al Hendrickson, Mike Zeman, Jack Ten Pas, Craig Enders, Peter Bemis and Ken Thiel.

The ADF approach set Bemis apart from thousands of commodity injection molders across the country. In part, because of ADF Bemis could offer customers much more—in engineering expertise, product design and successful molding of intricate parts for rigorous applications.

It was during these years that Bemis began to forge strategic partnerships with General Electric, DuPont, Dow, Geon and other suppliers of sophisticated new polymers. Industry alliances also included equipment manufacturers Cincinnati Milacron, Davis-Standard, and ABB.

These partnerships led to more opportunities for Bemis to take on challenging projects for high-profile industry leaders such as John Deere, Whirlpool, and Procter & Gamble. In turn, those projects led to greater national prominence for Bemis—and a greater competitive advantage in a growing industry.

A sophisticated multi-axis robot removes a resin loveseat from an injection molding press at the Bemis ADF plant in Sheboygan Falls. Built in 1988 the facility incorporated the company's Automated Defect-Free philosophy in a state-of-the-art manufacturing plant.

ADF technology spurs growth and expansion

In spring 1988 Bemis broke ground for the first phase of an ADF plant in Sheboygan Falls. The new facility was far from just another manufacturing plant. It was an Automated Defect-Free plant—designed to join the latest in process automation with employee empowerment and the Bemis philosophy of defect-free manufacturing.

Up and running in fall 1988, ADF was a state-of-the-art injection molding facility that brought people and technology together in a way that Bemis never had before.

Computers monitored production. Multi-axis robots removed parts from presses and performed secondary operations. Conveyors carried parts to operators who used statistical process control to check parts before assembly or packing. It was the future of manufacturing—and with ADF, that future had arrived at Bemis.

Lawn Comfort Division marks return to furniture roots

As Bemis expanded its large press molding capability, the company looked to make the most of that newly acquired technological expertise by considering new opportunities. One of these was in the furniture industry.

Company management had received word from European sales reps that demand for resin outdoor furniture was growing overseas and was expected to gain popularity in the United States. In 1988 Bemis entered this newly emerging market with the purchase of Lawn Comfort, U.S.A.

Based in Miami, Florida, the firm was the U.S. distributor for Lawn Comfort S.A. of Belgium—one of the leading outdoor resin furniture manufacturers in Europe.

To manage the new furniture product line at Bemis, Belgian native Philippe M. Ubaghs was hired as national sales manager. Philippe and the Miami operation were relocated to Sheboygan Falls, where Bemis set about applying its expertise in large part injection molding to resin furniture manufacturing. By 1989 Bemis had launched a new line of resin outdoor furniture and accessories to U.S. and Canadian retailers under the Lawn Comfort and Festival brands.

The acquisition was a return to the company's roots as a furniture manufacturer, dating back to the 1920s when Al Bemis owned a fibre furniture firm. Al had built that business on lessons learned from his father, Francis Tower Bemis, a nineteenth-century pioneer in Sheboygan County's early furniture-making trade. Now, over a century after his death, Francis' descendants Dick and Peter were returning to the industry where their great-grandfather made his mark.

European seat sales strengthened with first international acquisition

The purchase of Church Seats in 1983 had substantially broadened the company's core toilet seat business domestically. As the 1980s drew to a close, Bemis looked to further expand in Europe—this time through an acquisition outside of the United States.

British plumbing products distributor Harrison/Cavalier marketed Bemis seats throughout the United Kingdom under the Cavalier brand. A family-owned firm that had been in the sanitary plumbing business for five generations,Harrison/Cavalier was the exclusive distributor of Bemis seats to wholesale and retail channels in the U.K.

The Harrison/Cavalier plant in Silsden, West Yorkshire, England, became part of Bemis in 1987. The purchase was the first international acquisition in the 86-year history of Bemis and gave the company a base of operations in Europe.

When owners David, Fred and Stuart Harrison learned that Bemis was interested in purchasing a Euro-based seats manufacturer, they offered their own firm for sale in lieu of losing their primary supplier of toilet seats. Bemis acquired a majority interest in Harrison/Cavalier in summer 1987. It was the first international acquisition for Bemis, one that would give the company a foothold overseas—a base of operations to increase sales in markets throughout Europe. The Harrison/Cavalier acquisition signaled the official start of Bemis international operations and formed an early foundation for the company's growth in global markets.

In 1993 Bemis opened a new office expansion at Bemis-Cavalier in England. The new facility was part of the company's efforts to centralize its European operations. Celebrating here are (from left) Mayor of Burnley John Greenwood, Mayoress Margaret Greenwood, Bemis Senior Vice President of Finance Pete Lukaszewicz, Purchasing Director Stuart Harrison, Accountant Stephen Lane, Sales Manager Kieron O'Grady, Marketing Manager Vesla Bemis and Managing Director David Harrison.

Purchase of Moldex strengthens North American seat sales

Global growth through acquisition was something that Bemis had begun with the 1987 purchase of Harrison/Cavalier. Next stop on the map? Canada.

"Born in Barrie, Raised Everywhere"

"Born in Barrie, Raised Everywhere" was the clever tagline for Moldex, Ltd., Canada's largest toilet seat company. Based in Barrie, Ontario, north of Toronto, Moldex manufactured polypropylene, compression molded and soft vinyl toilet seats primarily for retail distribution in Canada.

In spring 1989 Bemis management including Corporate Legal Counsel John Howell, Pete Lukaszewicz in finance and Dick Bemis struck a deal to purchase Moldex. Within months the Canadian compression molding operation was relocated to Sheboygan Falls, along with Moldex faux woodgrain finishing machines.

After acquiring Canada's leading toilet seat manufacturer in 1989, Bemis retained the Moldex brand for some years to leverage seat sales in markets north of the U.S.

By the end of 1989 the facility was producing plastic and soft toilet seats under the Moldex brand name and resin furniture for the Lawn Comfort Division. With this move Bemis had officially established a manufacturing presence and a concerted sales effort in Canada.

Décor acquisition further broadens product mix

By 1993 Bemis had established a diverse product mix that spanned a host of industries from plumbing products to outdoor living, home improvement to housewares and health care to home comfort.

The company further diversified its product lines in January 1994 when Bemis management, led by Financial Senior VP Peter Lukaszewicz and VP of Sales and Marketing Dale Schmidt, finalized the purchase of Décor America. Based in Compton, California, Décor was the U.S. subsidiary of an Australian plastics manufacturer of planters, picnicware and kitchenware. With the purchase Bemis acquired tooling, products, patents and trademarks. In less than a year Décor offices and warehouse in California were vacated, and all operations relocated to Sheboygan Falls.

Décor was a natural fit for Bemis. Its unique self-watering planters complemented the Bemis casual outdoor furniture line. Décor plastic picnicware and kitchenware broadened the Bemis cutting board and cutlery product lines. The acquisition also signaled the start of yet another Bemis consumer products division—Gardenware.

The purchase of Decor in 1994 brought a new product line to Bemis—garden planters and pots.

Following the Décor purchase the acquisitions that drove growth at Bemis for over two decades slowed. From the company's acquisition of Wisconsin Plastics and formation of Bemis Health Care in 1971 to the purchase of toilet seat competitors Stasco, Church, and Moldex to expansion into new product categories through West Bend, Gamco, Lawn Comfort, and Décor, Bemis had grown at a rapid pace. The challenge was to manage that growth and channel it to expand market share in the industries of which Bemis was now a part.

The Bemis corporate logo graces the prominent entrance sign a the company's world headquarters in Sheboygan Falls, Wiscons This logo was introduced in 1996 as Bemis unveiled a broad global branding strategy that placed renewed emphasis on the company's flagship brand name.

CHAPTER 8

On the threshold of a new century

By the mid-1990s Bemis had positioned itself as a manufacturer and marketer of diverse consumer and health care products, and a world-class processor of engineered thermoplastics.

Growth and acquisitions over the past two decades had expanded company staff from 417 in 1975 to 1,796 employees in 1995. Multiple consumer brands included Lawn Comfort, Mayfair, Rain Master, Cavalier, Tommer, Colonial Kitchen, Waterwick and Décor. International sales were growing with the acquisitions of companies in Europe and expansion of market share worldwide.

Together these changes transformed the Bemis of the 1990s into a far different company than the small Wisconsin wagon and woodworking firm founded in 1901.

But the world of business and technology had undergone much change in those 90 years. A global economy was becoming more prevalent as trade barriers crumbled. High-speed communication was fast-forwarding technology into the next century with cellular phones, laptop computers and a worldwide web of information. As Bemis rolled toward the new millennium, the company took full advantage of the opportunities presented by these cultural and industrial shifts—and expanded its technology, customer base and success in the process.

President and CEO Dick Bemis (seated) enjoys the comfort of a company-made resin chair while Philippe Ubaghs, vice president of the Bemis Casual Living Products Group, looks on.

Furniture flourishes

The mid-1990s were the heyday for casual furniture production at Bemis. New materials, colors, technology and styles all contributed to the company's success in this category as Americans embraced outdoor living and sought carefree products to furnish patios, decks and gazebos. Resin outdoor casual furniture offered it all—light weight, unique design, easy storage, low maintenance and moderate price.

Bemis resin casual furniture was offered in collections to help homeowners select pieces that coordinated with one another. The Bemis Gallery collections also helped the company's retail partners increase sales and profits.

With its large capacity injection molding presses, robotics and advanced process technology Bemis was well-positioned to manufacture resin tables, chairs, chaise lounges, beverage carts and more. Combined with the company's new foray into gardenware, kitchen storage containers and plastic dinnerware with the 1994 acquisition of Décor America, Bemis could now offer consumers a host of colorful plastic products for indoor and outdoor living spaces.

New logo, new look, new branding effort

Mayfair, Moldex, Church, Cavalier, Bemis—all were part of the stable of toilet seat brands that Bemis promoted in the late 1980s and early 1990s. But these multiple brands, while distinguishing Bemis products in different markets, were difficult to manage. More product lines, more manufacturing, more packaging and more inventory all meant more bottom-line cost to Bemis—and sometimes more confusion in the marketplace.

By the mid-1990s Bemis recognized the value of moving toward fewer brands and using the Bemis name as the company's primary brand for recognition, internal efficiencies and a more unified marketing effort.

So in 1996 Bemis adopted a modernized logo—the fifth in the company's near-100-year history. This dynamic redesigned logo embodied the Bemis twenty-first-century spirit by reflecting the company's solid foundation and strong sense of forward movement. Presented in bold Bemis blue, the logo also communicated innovation—a company cornerstone for 100 years.

The diversified consumer products company that Bemis had become by the mid-1990s is reflected in the photo above. From humidifiers to kitchen cutting boards and rain gutters to resin casual furniture, Bemis manufactured diverse products for homeowners.

1994	1994	1995	1996	1998	1999	1999	2000
Millions watch the fall of American hero O.J. Simpson in a televised chase on California freeways.	Bemis opens a first-ever facility in Mexico—a sales office in Monterrey.	The worst act of terrorism in U.S. history rocks the nation as a car bomb rips a nine-story hole in the Oklahoma City Federal Building.	Bemis forges a joint venture with a Chinese injection molder to supply toilet seats to the Pacific Rim.	The Bemis acquisition of Kelch Corporation is the largest in the company's history.	Computer programmers worldwide look to stamp out electronic bugs as the year 2000 approaches.	A new finishing line is added at Plant B to boost capacity and support domestic and international sales gains.	Genetic researchers assemble the first working draft of the human genome, a code book for human life.

Dual brand strategy spells seat success

The company's new branding strategy included retaining two toilet seat brands for the wholesale channel—Bemis and Church. This dual trade brand approach was first initiated by seats National Sales Manager Don Wilkens and was later advanced by his successor Jim Roenitz.

Under Jim's direction the Church line was injected with new life in the mid-1990s, including new products and features, a new logo and product literature, new sales reps, national trade advertising campaigns and promotions tied to the brand's 100th anniversary in 1998.

The efforts paid off. Market share increased significantly, and today Church is solidly positioned as the premium choice for the plumbing wholesale channel—true to the brand's "Best Seat in the House" tagline.

In the mid-1990s Bemis positioned the Church name as an exclusive brand available only at wholesale. Part of the company's brand strategy for seats, the Church marketing effort included trade ads like the one below.

Seat innovations drive sales

During the mid- to late-1990s Bemis introduced a host of toilet seat innovations. Pivotal among those was the Bemis Proseat.

Featuring the company's patented top-tightening UltraHinge, a rock-hard three-layer enamel finish and a sleek new design, this molded wood seat offered plumbing professionals more durability, style and value than ever before. The Proseat quickly became an industry standard and today remains the molded wood seat by which all other seats are judged.

The UltraHinge wasn't the only Bemis hinging innovation to make waves in the industry in recent years. At the 2000 National Kitchen and Bath Show Bemis introduced a unique slow-close, antislam hinge to protect the seat, china toilet bowl and user. The action of raising the seat lid activates the slow-close hinge so just the touch of a finger starts the seat on its gentle, gravity-driven descent.

Bemis took hinging to even greater heights with the 2001 introduction of its easy-to-clean lift-off toilet seat. The unique hinge attaches the seat securely to the bowl, yet easily releases it for cleaning. These and other innovations in toilet seat design mirrored consumer demand for ease of use and high style. They also demonstrated the company's long-standing industry leadership and the unique products that supported that leadership for a century.

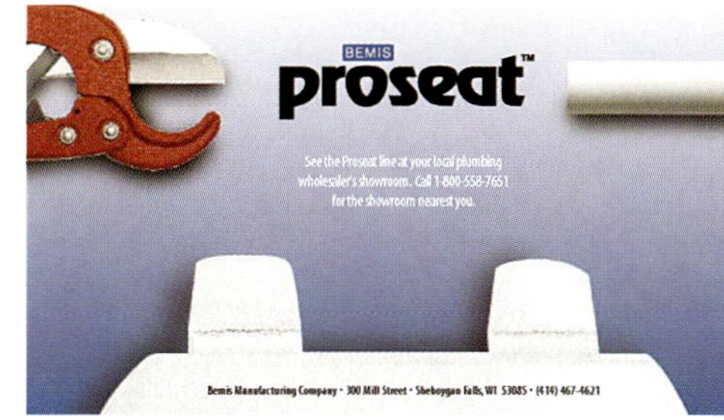

Bemis provided the right tool for the right job to professional plumbers with its Proseat, which debuted in 1994. The seat's unique top-tightening UltraHinge, combined with its durable three-coat finish and sleek style, helped solidify the company's market leadership position.

Bemis used this humorous image of a spike-collared bulldog grasping a toilet seat in its jaws to illustrate the durability of its enamel paint finish. The patented Bemis three-layer finish, applied to all the company's Molcor toilet seats, is rock-hard, water-based, flawless in appearance—and leaves the competition chasing their tails.

Strategic leadership

Bemis surely wouldn't be the successful, stable company it is today were it not for the vital roles played by key corporate executives in the financial and human resources areas.

Peter Lukaszewicz, who passed away in 2000, and his predecessor, James Kroscher, elevated their roles as CFO from simply reporting financial results to that of strategic partners with the Bemis family and the company. Pete and Jim worked to ensure that pricing and profit margins were adequate to maintain the company's success.

Pete Lukaszewicz

After the death of F.K. Bemis in 1987 Pete also provided an essential objective viewpoint on business decisions for F.K.'s sons, Dick and Peter.

Jim Kroscher

And Jim's influence continues to be felt as a member of the Bemis Board of Directors, a position which he assumed after his retirement from the company in 1986.

In the Human Resources arena, Ed Collins brought wisdom, insight and stability to the company's hiring practices and HR functions. Ed retired from Bemis in 1999 as vice president of Human Resources.

Ed Collins

Together, Ed, Pete and Jim took a proactive approach to managing the the business at large. Their contributions have helped Bemis achieve its current success and position the company securely to meet the challenges of the next century.

Contract sets new standards with co-injection

As plastic applications continued to evolve with newer and better materials, companies like Bemis were handed the challenge of processing them effectively. Creative design, advanced engineering and new technology were the tools that Bemis used to broaden its reputation and customer base in contract molding and extrusion.

One of the key technologies the company used was co-injection molding. A cost-efficient process for forming complex plastic parts, co-injection molding overlays an inexpensive material with a more costly prime-grade plastic. The result was a high-performance plastic part with unique properties and top aesthetics—at reduced cost.

Bemis technologies like co-injection molding, quick color changes and large press molding set the company's Contract Division apart from thousands of low-cost commodity plastics processors who focused on price alone. Technology was the ticket that propelled Bemis into an elite top tier of plastics processors, companies that could offer customers unique parts with unique shapes and properties—parts that drove the plastics industry forward.

Bemis garners success by association with Kelch

Since the mid-1970s Bemis Contract had been forging strategic alliances with key industry players including plastic resin makers, equipment manufacturers and other suppliers. The goal? To help OEM customers bring better products to market in less time and for less cost.

One of these partnerships was with a Wisconsin-based maker of fuel containment systems for non-automotive applications. That company was Kelch Corporation.

Bemis and Kelch first made one

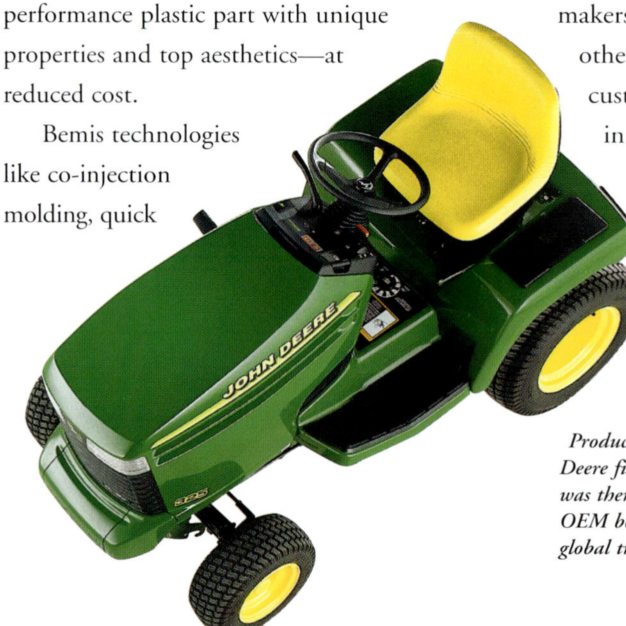

Product development support of mutual customer John Deere first brought Bemis and Kelch together in 1993. It was then that the two companies were manufacturing OEM body and fuel system components for the leading global tractor and lawn and garden manufacturer.

another's acquaintance in 1993 as key suppliers to tractor industry giant Deere & Company. Like many companies, John Deere sought to reduce its vendor base, decrease the number of parts in its tractors, and leverage recent advances in materials engineering and supplier technology, all while enhancing aesthetics and performance of its products. With that formidable challenge at hand, Bemis and Kelch formed a collaborative alliance of OEM suppliers to offer John Deere and other customers synergistic new product development assistance. Named "The Consortium," the Bemis/Kelch alliance leveraged their relationship to make joint presentations to customers, engage in technology exchanges and share best practices—all in an effort to help one another win in the marketplace.

In spring 1997 the companies launched a joint-venture expansion at a Kelch satellite manufacturing facility in Lenoir, North Carolina. Strategically located in the southeastern United States—a major manufacturing hub for lawn and garden equipment —the Lenoir facility offered Bemis a base of operations to service rapidly growing area industries.

Consortium leads to Kelch acquisition

While small part injection molding was not new to Bemis, the specialized fuel system components that Kelch manufactured were.

In the 1960s and 1970s, Bemis Contract operations had molded a host of small parts on small presses for a variety of OEM customers nationwide. But by the 1980s the company began moving from small part molding to large parts such as tractor hoods and body panels for the lawn and garden industry. This migration to large part molding culminated in July 2000 with the arrival of a mammoth 6,600-ton co-injection press at ADF facilities in Sheboygan Falls—the largest co-injection press of its kind in the world.

As Bemis made these and other strides in large part molding, Kelch carved out a successful niche in small part molding of highly specialized precision fuel containment systems for off-road applications.

Bemis and Kelch both recognized they had much in common—complementary capabilities and press sizes, common OEM customers and markets, and comparable commitments to technology, employees and the communities in which they resided. So in late 1998 Bemis took the next logical step in evolving its long-standing relationship with this strategic partner and purchased Kelch Corporation. It was the largest acquisition

The largest machine of its kind in the world, this mammoth 6,600-ton Milacron Maxima co-injection press arrived at Bemis ADF facilities in mid-2000. The state-of-the-art behemoth has a footprint the size of a three-bedroom house and a height that soars to 18 feet. An adult man, shown on the operator platform (in red circle) near the center of this photo, is dwarfed in comparison to this enormous press.

With the 1999 corporate reorganization Peter F. Bemis was named as president of the new Contract Products Group.

in Bemis history, bringing together two powerhouse suppliers for the nonautomotive industry to create new opportunities for both companies—and for the customers they served.

The Kelch purchase placed Bemis among the top ten injection molders for nonautomotive markets in the United States. It also brought a new manager to the company's executive leadership with the addition of Kelch president Dennis Nourse.

Reorganization signals new era at Bemis

By the late 1990s rapidly shifting markets, escalating customer demands and widespread globalization were putting pressure on Bemis. More technology, faster delivery, lower cost—all were mandated if Bemis was to thrive in the twenty-first century.

To meet these realities, company CEO Dick Bemis announced a corporate reorganization in summer 1999. This strategic realignment created two distinct business groups—Contract and Proprietary. Peter F. Bemis was named president of the Contract Products Group, and former Kelch President Dennis Nourse was chosen to lead the company's Proprietary Products Group.

Former Kelch President Dennis Nourse was chosen to lead the newly formed Proprietary Products Group as part of the company's 1999 reorganization.

The 1999 reorganization was a noteworthy event in the company's transition from a medium-sized domestic-based plastics manufacturer to a major supplier of products for a host of global industries.

Bemis subsidiary Kelch Corporation offers the world's best selection of caps, gauges and sensors for nonautomotive applications. Kelch also manufactures steering wheels and fuel tanks for off-road use and markets these highly specialized parts to tractor and lawn and garden OEMs worldwide.

Bemis bids farewell to Plant A

Dating back to the company's earliest years, Bemis Plant A in Sheboygan Falls was razed in late 1999 to make room for Settlers Park.

It was the birthplace of Bemis Manufacturing Company. An aged building in ill repair on the west bank of the Sheboygan River just outside of downtown Sheboygan Falls.

The Bemis Broadway Avenue plant—or Plant A as it had come to be known over the years—had reached the end of its useful life by summer 1999. While the company had retained a small finishing line on its second floor, the majority of Plant A stood dusty and vacant, symbolizing the Bemis migration to newer technologies and facilities to meet modern marketplace demands.

Realizing that the value of Plant A was its prime riverfront location, Bemis chose to close the facility, raze it and donate the land to the city of Sheboygan Falls for Settlers Park, a scenic public recreation area near the upper dam of the Sheboygan River.

Prior to demolition Bemis hosted an outdoor celebration at Plant A. The August 1999 party paid tribute to the building that housed corporate headquarters from the 1930s to the 1960s and manufacturing operations for nearly 70 years. It also honored hundreds of employees who worked at Plant A during those vital years in Bemis history.

Dick Bemis (right) reminisces with Edna (Mickey) Kohlhagen, a Bemis employee for 57 years, and Peter Van Wyk (left), a former employee of Bemis.

Bemis gives back to the community

Bemis has had a presence in Sheboygan Falls since the 1864 arrival of Francis Tower Bemis, great-grandfather of company leaders Richard A. and Peter F. Bemis. In April 1995 that presence was strengthened with the opening of the Bemis Home & Garden Factory Direct Store.

Occupying the restored James W. Smith building in downtown Sheboygan Falls, the Home & Garden Store features toilet seats and other consumer home products manufactured by Bemis.

Originally constructed in 1880 as a meat market and post office, the Smith building was purchased in late 1994 by Bemis, which supervised its historic renovation. Listed on the National Register of Historic Places and recognized as one of the most noteworthy Italianate commercial buildings in Wisconsin, the Home & Garden Store continues to be a vital part of the rebirth of the downtown historic district in Sheboygan Falls.

Along with the company's 1999 donation of the former Plant A property for Settlers Park, the Home & Garden Store embodies the Bemis commitment to give back to local communities where it resides. Responsible corporate citizenship is also reflected in the company's generous support of educational, health care and other nonprofit organizations through The F.K. Bemis Family Foundation.

The James W. Smith building in downtown Sheboygan Falls once housed a meat market and post office. Today it is occupied by the Bemis Home & Garden Factory Direct Store.

Mayfair, a Bemis brand for the retail sales channel, was first introduced to the Mexican marketplace in fall 2000. Colorful trade booth graphics such as this were displayed at the Bemis booth at Expo Nacional Ferretera, a major Mexican trade show for the retail hardware/DIY industry.

South-of-the-border seats expansion

Crumbling trade barriers, common currencies and industry consolidation in the mid- to late-1990s all signaled the growth of a global marketplace, one where Bemis could command a leadership position in toilet seat manufacturing and marketing.

The company's 1988 acquisition of Euro-based Harrison/Cavalier was a first step toward this world leadership. The purchase of Canadian seats leader Moldex in 1989 was a second move toward international growth. In 1994 south-of-the-border sales were set in place with the opening of a first Bemis presence in Mexico.

The business began with a one-man sales office to service Mexican retailers, wholesalers and OEMs with Bemis toilet seats manufactured in Sheboygan Falls. In time the operation expanded, adding new products, facilities and staff to gain market share.

By late 1999 Bemis completed construction of an impressive 40,000-square-foot facility near Monterrey that included manufacturing, warehousing and office space. It was a prime location to leverage the company's toilet seat line and offer local sales, service and support—a must for Bemis to capture market share in Latin America. The Monterrey facility proved to be a strategic base of operations for additional expansion into South America as Bemis established offices in Chile and Brazil.

Rich, vibrant hues are the colors of choice for toilet seats in many overseas markets. As Bemis expanded sales and marketing in Latin America, Europe and beyond, the company developed new seat shades and shapes to meet unique preferences in markets across the globe.

European presence strengthened by Italian acquisitions

One of the challenges Bemis faced as it considered new overseas markets was matching its product line selection to the preferences of foreign customers.

No one knew this better than Philippe Ubaghs. A native of Belgium, Philippe came to Bemis as part of the Lawn Comfort acquisition in 1988. A decade later he was promoted to lead Bemis European operations.

Philippe and his overseas staff recognized that while Americans prefer toilet seats in white or natural colors, more vibrant hues were preferred in other countries. Overseas markets also favored different designs, materials and dimensions.

While Bemis could adapt its color and style offering to match local market needs, the company found it could not as easily offer plastic products in melamine or polyester, the materials preferred by Europeans. Bemis plastic seats were molded of polypropylene, and the company had neither the manufacturing equipment nor the technical expertise to manufacture seats in melamine or polyester.

Rather than allocate time and money to develop that expertise, Bemis chose to purchase this knowledge and manufacturing capability with the December 1999 acquisition of Carrara & Matta.

A leading Italian plumbing products manufacturer, Carrara & Matta was proficient in melamine toilet seat manufacturing and had industry connections to subcontract production of polyester seats. With the acquisition Bemis became the largest seat manufacturer in Italy.

LAST EXPORT

Within months of the Carrara & Matta purchase, Bemis acquired a second Italian firm—PlastExport, one of Europe's top toilet seat manufacturers for the OEM distribution channel. The purchases of Carrara & Matta and PlastExport significantly strengthened the company's European presence and its commitment to be a full-line supplier to markets worldwide. The Italian acquisitions also broadened the Bemis customer base to include new European firms. In addition, they enhanced Bemis support of domestic customers, many of which had rapidly expanding global operations. Together, Bemis and its European subsidiaries were well-positioned to grow OEM, retail and wholesale seat sales internationally as the company marked a century in business in 2001.

CARRARA & MATTA

A unique and stylish product line of Italian designer toilet seats and bathroom accessories came to Bemis with the 1999 acquisition of Carrara & Matta. This European plumbing fixtures manufacturer brought Bemis new technology and products for new applications in new markets worldwide.

The Bemis executive management team was photographed in January 2001 at the inception of the company's 100th anniversary. They are (from left):

FRONT ROW

Steven E. Larson
Vice President, Market & Business Development, Contract

Peter F. Bemis
Executive Vice President, Bemis Manufacturing Company
President, Contract

Richard A. Bemis
President and CEO, Bemis Manufacturing Company

Dennis E. Nourse
President, Proprietary Products

James M. Roenitz
Vice President of Marketing, Proprietary Products

Thomas R. Weber
Vice President of Sales, Contract

Bruce J. Hagen
Vice President, Corporate Information Systems

BACK ROW

John B. Howell
Corporate General Counsel

Dale E. Schmidt
Vice President of Sales, Proprietary Products

Bill Foster
President, Kelch Corporation

John E. Rummele
Vice President of Engineering, Proprietary Products

*Not shown is **Philippe M. Ubaghs,** Vice President and Managing Director, Europe*

Bemis at its 100th year in business

The extraordinary growth that Bemis has experienced from 1901 to 2001 is evidence of the company's century of success.

Today Bemis has global operations on four continents. The company employs over 3,000 people at over 2.8 million square feet of manufacturing, distribution and office space worldwide. Annual sales reached a record high in 2000.

As it looks ahead to the next century, Bemis continues to focus on its areas of expertise. Foremost is a commitment to become the premier supplier of toilet seats in the world. To secure and retain global leadership, Bemis has committed significant capital investment to expand facilities and staff and increase seats market share and penetration worldwide.

Another core business unit is the Contract Products Group. Bemis has made a major commitment to supply North American industries with specialized and highly engineered thermoplastic components. Technology-driven and process-focused, Bemis Contract Group offers innovative manufacturing in impressive state-of-the-art facilities. The operating unit plays a key role in a host of industries including lawn and garden, large appliance, electronics and office systems.

Through its subsidiary, Kelch Corporation, Bemis has directed substantial resources to gain North American market leadership in fuel containment systems for nonautomotive applications. The specialized gas caps, tanks, gauges, sensors and other products manufactured by Kelch make this Bemis subsidiary a key supplier to nonautomotive markets worldwide.

As Bemis marks its 100th year in business in 2001, the company's management team looks to the future and the roles that fourth-generation family members may play. Comprised of the children of Dick and Peter Bemis, this generation is poised to contribute to their family's century-old firm by maintaining its private ownership status. Other contributions from the next generation are yet to come.

The Bemis Board of Directors, photographed in May 2001. They are (from left): Robert Melzer, James Schreiber, Peter Bemis, Jeffrey Trowbridge, Richard Bemis and James Kroscher.

In the meantime, Bemis Manufacturing Company remains a strong, viable, financially stable corporation at its centennial year—privately held, multinational and well-positioned for future growth and success.

Global Facilities

Bemis Manufacturing Company

Corporate Headquarters
Sheboygan Falls, Wisconsin

Bemis Southeast
Lenoir, North Carolina

The Kelch Corporation
Cedarburg, Wisconsin

Bemis de Mexico
Monterrey, Mexico

Bemis Canada
Barrie, Ontario

West Campus
Sheboygan Falls, Wisconsin

Bemis
Menomonee Falls, Wisconsin

Bemis
Crandon, Wisconsin

Bemis Ltd.
Burnley, Lancashire, England

Bemis S.r.l.
Volpiano (Turin), Italy

Index

Page numbers in *italics* refer to illustrations or photos and their captions. Page numbers followed by "t," for example 3t, refer to information in timelines.